Dynamic fracture

Dynamic fracture

edited by

M.L. WILLIAMS
University of Pittsburgh
Pittsburgh, Pennsylvania
USA

and

W.G. KNAUSS
California Institute of Technology
Passadena, California
USA

1985 **MARTINUS NIJHOFF PUBLISHERS**
a member of the KLUWER ACADEMIC PUBLISHERS GROUP
DORDRECHT / BOSTON / LANCASTER

Distributors

for the United States and Canada: Kluwer Academic Publishers, 190 Old Derby Street, Hingham, MA 02043, USA
for the UK and Ireland: Kluwer Academic Publishers, MTP Press Limited, Falcon House, Queen Square, Lancaster LA1 1RN, UK
for all other countries: Kluwer Academic Publishers Group, Distribution Center, P.O. Box 322, 3300 AH Dordrecht, The Netherlands

Library of Congress Cataloging in Publication Data

Main entry under title:

Dynamic fracture.

1. Fracture mechanics--Addresses, essays, lectures.
2. Structural dynamics--Addresses, essays, lectures.
I. Williams, M. L. (Max L.) II. Knauss, Wolfgang Gustav.
TA409.D96 1985 620.1'126 85-10582
ISBN 90-247-3179-8

ISBN 90-247-3179-8

PRINTED IN THE NETHERLANDS

CONTENTS

Foreword

From time to time the *International Journal of Fracture* has presented matters thought to be of special interest to its readers. In previous special issues (December 1980 and April 1981), Dr H.W. Liu as Guest Editor presented a series of review papers dealing with fatigue processes and characteristics in metals and non-metals. Continuing this policy, which is consistent with our stated objectives, a second review dealing with time dependence in the fracture process, including the effect of material inertia but essentially excluding very strong shock effects in solids, has been assembled under the generic term "dynamic fracture". We hope that the ensuing state-of-the-art review will yield an instructive and timely product which readers will find useful.

To assist us in presenting this subject, we have prevailed upon a well-known worker in dynamic fracture, Dr W.G. Knauss, Professor of Aeronautics and Applied Mechanics, California Institute of Technology to act as Guest Editor for this special double issue. On behalf of the editors and publisher, I wish to express our indebtedness to Professor Knauss and his invited authors for undertaking this special effort.

M.L. Williams
Editor-in-Chief

Pittsburgh, Pennsylvania
March 1985

Preface

W.G. KNAUSS

During the past three decades significant studies have been made in understanding fracture mechanics. These advances have encompassed both the fracture process itself and improved design criteria for engineering structures. Failure prevention in pressure vessels or in aerospace structures would be unthinkable today without fracture mechanics principles. Similarly, the development, improvement, and evaluation of materials depends upon many of the same advances.

The majority of past work on fracture has been devoted to the events which lead up to the onset of crack propagation. However, it is equally important to be concerned with those processes that govern the time dependent motion or growth of cracks. Time dependence in the fracture process develops for various reasons, such as viscoelastic or viscoplastic material behavior, cyclic loading histories leading to the phenomenon of fatigue, or through the involvement of material inertia in the form of stress waves. While these problems span a spectrum of time scales, it has been the practice to distinguish between situations in which the action of stress or shock waves dominates the fracture process and those in which material inertia plays a less significant role. Viscoelastic or creep fracture and fatigue fall in the latter category, although dynamically induced vibrations are often the cause of fatigue failure.

In this book the authors will review certain aspects of dynamic fracture in which stress waves dominate the crack propagation processes. This distinction is, however, not always clear cut and can include a spectrum of situations with inertia related vibrations, up to and including the effect of shock and impact physics. Such latter situations usually involve extensive fracturing. For this book, however, we chose to emphasize those problems in which individual cracks play a dominant role.

Systematic investigation of dynamic fracture and crack propagation commenced with Schardin's exploratory but extensive experiments in the 1930s and continued with Rinehart and Pearson's spall and impact studies about a decade and a half later. Since then interest and sophistication in dynamic fracture has grown at an accelerating pace, motivated by several important engineering problems. They include: The safety of nuclear pressure vessels under rapid cool down which can spawn dynamically growing cracks; The rapid propagation of cracks in gas and off-shore oil pipelines can generate economical problems of large proportions; Dynamic comminution arises in the pharmaceutical industry as well as with explosively generated porosity in geological formations for augmented gas and oil recovery; Nucleation and growth of micro-cracks in the earliest stages of fatigue failure.

The diversity of these problems is an indication of the breadth of topics considered under the heading of Dynamic Fracture. A full exposition is not possible, and indeed is not intended, for the immediate review purpose. Nevertheless, it was deemed desirable to collect a number of papers, each with a large component of review material, to summarize at least a significant portion of this field. It is hoped that such a "snapshot" of the current state of affairs will illuminate common underlying principles and behavior in such varied situations.

Much of the work of the last decade was concerned with the experimental definition of the state of stress in laboratory specimens for which analytical solutions could not be obtained. That work is not represented here in proportion to the effort expended on it. The applications of results to dynamic fracture in reactor vessels and pipelines is also not represented extensively. Fracture in geological formations, as occurs in earthquake related problems, is only presented in one paper to present a frame of reference to exhibit similarities and differences commended to the reader's attention. Finally, and despite the lack of emphasis on gross spallation, there are presented some problems of crack initiation for single and multiple cracks which might be applied to an elemental understanding of the general spall problem. In this characterization, the stress field around the moving crack tip in model and actual structural materials are considered from both the experimental and analytical point of view. Finally the question of dynamic crack branching which governs the technical problem of comminution receives considerable attention as does the process of how cracks come to rest under dynamic loading.

The reader will find that many questions on these topics remain unanswered. We do hope however that someone who has a basic appreciation of fracture problems can readily evaluate the current state of knowledge in this field.

W.G. Knauss
Guest Editor

Pasadena, California
March 1985

International Journal of Fracture 27 (1985) 127–143.

Some basic problems in stress wave dominated fracture

W.G. KNAUSS * and K. RAVI-CHANDAR **

* *Graduate Aeronautical Laboratories, California Institute of Technology, 105-50, Pasadena, CA 91125, USA*
** *Department of Mechanical Engineering, University of Houston, Houston, TX 77004, USA*

(Received September 1, 1984)

Abstract

Problems of characterizing the stress wave induced stress field around the crack tip and ensuing crack propagation behavior are described. After stating the assumptions underlying an idealized model of brittle, dynamic fracture we review experimentally observed deviations from such idealized behavior. Such deviations are ascribed to the development of numerous microcracks in the crack tip region. These microcracks determine the crack initiation process, the speed of crack growth and the branching phenomenon. Also we examine critically the contention that there exists a unique relation which links instantaneous crack speed to the instantaneous stress intensity factor in brittle solids. That question hinges materially on limitations of current measurement techniques.

1. Introduction

The search for criteria of material strength is an old development which spawned the theory of elasticity and plasticity. Yet when the latter had developed to the point where useful comparisons between simple failure (fracture) criteria and experiments were feasible it became clear that merely knowing the state of stress throughout a solid may not be sufficient to characterize fracture or even its initiation. It was primarily out of this recognition that the quest for a more clearly physical failure process led to the study of cracks and their role in failure. The introduction of preexisting micro cracks or other flaws into failure considerations provided a new physical basis for better scientific investigations into material strength.

With the flaw concept in mind essentially two areas of investigation were opened: One avenue leads to the characterization of strength through a dominant, macroscopic crack while the other follows the development and growth of multiple but more microscopic flaws. In the latter category we find the problems of explosive comminution and spalling such as are prevalent in the investigation of the Hopkinsons [1,2], Rinehart [3], Kolsky [4], and Curran, Shockey, and Seaman [5]. Of primary concern in these investigations is the more or less simultaneous development and growth of many small flaws which later coalesce and thus generate a macroscopic failure. The most recent and to date most detailed investigation of the physical processes involved in spall development is due to Curran, Shockey, and Seaman [5] who draw, for modelling purposes, on related ideas of rate dependent flaw growth by Zhurkov and colleagues [6,7].

Apart from this more spatially diffuse fracture process one finds that macroscopic crack behavior dominates investigations of dynamic fracture. Here one deals with the rapid growth of a single crack in response to static or rapid loading; most of the contributions in the double issue of the *International Journal of Fracture* on dynamic fracture are devoted to problems in this category. It is our current perception that the multiple flaw problem discussed above and this macroscopic crack problem are linked intimately, not only for the dynamic crack generation (spall) but also for the growth of macroscopic cracks. In this view the macroscopic crack provides primarily a highly localized stress and strain concentration associated with locally high strain rates to which the material at the crack tip responds by distributed micro fracture and/or hole growth. Thus the problem of dynamic growth of a macroscopic crack is really tied to the problem of multiple flaw growth which is normally investigated separately from the macroscopic crack problem. The major problem in connecting these "two problem areas" is that the realistic stress field in which the multiple flaws develop at the crack tip is, as yet, poorly defined. Furthermore, the statistical distribution of the micro flaws at the crack tip as well as the associated stress gradients make it difficult to apply growth models for the micro flaws. It is a partial purpose of this paper to illustrate the significance of this concept for at least the class of what we would like to call "quasi-brittle" materials.

This concept of a strong influence of many small cracks on macroscopic fracture does not conform to the currently "classical" approach to dynamic crack analysis and is, most likely responsible for many of the discrepancies between theoretical predictions and experimental observations. To begin with we shall, therefore, describe the idealized formulation of the dynamic crack growth problem and then proceed to examine its implications experimentally. Next, we shall discuss the experimental and analytical comparison of the crack tip stress field and then consider under the heading of nonideal fracture several discrepancies between experiments and the idealized view of dynamic fracture.

2. Idealized crack model

The presently dominant analytical model has been formulated after the static brittle fracture model of Griffith: One considers essentially rectilinear crack growth in the elastic plane such that the crack tip generates unbounded stresses with the fracture process occurring at the point of the crack tip proper. Moreover, one assumes that the expended energy per unit of new surface Γ, is a constant. These modelling assumptions determine the evaluation of the elastodynamic stress field at the crack tip and the formulation of an energy balance. The stress analysis for the idealized problem has been documented in various forms [8,9] and the energy balance analysis has been documented succinctly in [10] and [11]. The former establishes the square root singular crack tip behavior and its dependence of crack speed on the stress intensity factor ($K \to O$ as the crack speed $v \to C_R$, the Rayleigh wave speed).

The idealized modelling of fracture events would thus seem to require that initiation of crack motion, crack arrest as well as the crack propagation phase derives from the constancy of the fracture energy Γ. We know today that all these phenomena appear to be somewhat disconnected as evidenced by numerous "critical stress intensities" associated with crack start with arrest and crack propagation [12-I]. While researchers have thus, abandoned to a large extent, the pretexts of idealized fracture they do adhere to the concept of describing fracture events in terms of the stress intensity factor – which practice arises more out of necessity than physical reality. Clearly, the search is on for defining in a better way the physical processes that occur at the tip of a rapidly moving crack.

The knowledge of the idealized crack tip stress field and the assumption of a constant, i.e. rate and history independent fracture energy establishes per force a relation between the instantaneous stress intensity factor and the corresponding, instantaneous crack speed.

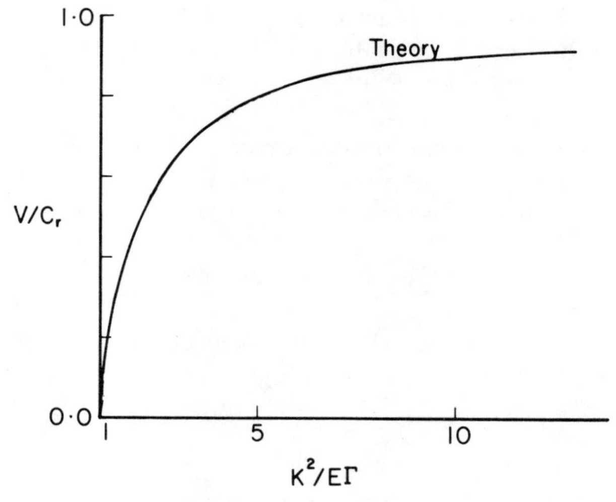

Figure 1. Relation between stress intensity factor and velocity for idealized brittle fracture.

Specifically, this energy balance equation becomes

$$\frac{E\Gamma}{K^2} = g(v) \doteq \left(1 - \frac{v}{C_R}\right) \tag{1}$$

where E is Young's modulus; for the definition of $g(v)$ see, e.g. [13]. From this relation one derives the "idealized" and unique relation between K and v as shown in Fig. 1 where the maximal crack propagation velocity is the Rayleigh surface crack speed. This idealized relation is not achieved in experiments, yet some researchers believe that a different, though nevertheless unique relation exists between the instantaneous stress intensity factor and the instantaneous crack velocity. We shall discuss this question in more detail later on, but it is clear that this is a topic of current research interest.

With regard to the phenomenon of branching the idealized model presents no automatic explanation. Therefore, one tries to extend it by adhering to the dissipation of energy in the stress singularity at the crack tip point and by reformulating the energy balance in terms of two cracks [14]. The consequence of that reformulation is that for the idealized model ($\Gamma = \text{const}$) the velocity of the branches has to be less than that of the crack before branching and that branching can occur only if the initial crack propagates at least with $v = 0.5\ C_R$. Both conditions are rarely met in experiments; specifically, branching does indeed occur when the crack speed is less than $0.5\ C_R$.

2.1. The dynamic crack tip stress field

It is fair to say that both analytically as well as experimentally oriented dynamic fracture work rests essentially on two-dimensional concepts in the elastic plane, although virtually all realistic problems generate three-dimensional stress fields. This is particularly true in the vicinity of the crack tip. Although it is generally assumed – by necessity – that the resulting approximations are not serious this contention clearly limits the exploration of the details of the fracture process to the more prominent rather than detailed features of fracture.

The early work in the experimental determination of dynamic crack tip stresses [15] rested, with one exception [16], on the use of the (quasi) static stress field characterization at the crack tip. The viewpoint that quasistatic fracture concepts and analyses are applicable to the stress wave controlled fracture process has persisted into the present (see, e.g. [17] and [18] even though quite a reasonable amount of understanding involving crack tips and wave action has become available [19–25]. Not only was and is the quasistatic stress field used for data reduction, analyses of quasistatic situations are substituted for dynamic events [26].

As stated before, the assumptions of idealization lead to a singular stress field; for a stationary crack tip subject to time varying boundary loads one has for the cartesian components of the stress tensor (in polar coordinates r, θ)

$$\sigma_{\alpha\beta} = \frac{K(t)}{\sqrt{2\pi r}} f_{\alpha\beta}(\theta) + 0(1) \qquad \alpha, \beta = 1, 2 \quad \text{as} \quad r \to 0 \tag{2}$$

which field is established roughly in the domain about the crack tip where the shear wave emanating from the crack tip has passed [27]. The corresponding relation for a crack propagating with a velocity v is

$$\sigma_{\alpha\beta} = \frac{K(t, v)}{\sqrt{2\pi r}} f_{\alpha\beta}(\theta, v) + 0(1) \qquad \alpha, \beta = 1, 2 \quad \text{as} \quad r \to 0 \tag{3}$$

where now r, and θ are polar coordinates centered at the moving crack tip. "Closed form" analytical results are available only for the infinite domain. Experiments are, however, usually conducted on small geometries. For this reason we know of no comparison between an analytical solution and experiment for dynamically moving cracks. The only direct comparison between an analytical characterization of dynamic crack tip stress field for a crack that transits from being stationary to moving [27] and the corresponding experiment has been reported in [28] for plates of Homalite 100 via the method of caustics. Upon taking account of wave dispersion in plates of finite thickness through use of a plate velocity in place of the dilatational wave speed the comparison, shown in Fig. 2, is quite satisfactory.

On the other hand, nearly identical experiments with the same material, except under higher rates of loading and with higher stresses resulted in agreement only up to the time of crack propagation (cf. Fig. 3). Whether such disagreement is due to possible violation of the assumptions underlying the idealized fracture model or whether it is due to an inadequate accounting of the time development of the crack tip stress field is presently under study [29,30]. When realistic materials are involved other phenomena may overshadow or alter the idealized stress field so that close correspondence may not be achieved. For example, if the fracture and crack tip unloading processes are accompanied by considerable "plastic" crack tip deformation the fracture wave signals at the crack tip are correlated over a finite domain at the crack tip and would involve both unloading and loading simultaneously thus altering the just discussed stress field history possibly significantly.

Regarding these uncertainties of the crack tip stress field we note that for Homalite 100 Freund [29] has computed the related times and found that it takes a surprisingly long time – tens of μsec – to establish the two-dimensional square-root-singular stress field in a region around a moving crack tip from which the stress intensity factor can be determined. For shorter times the stress field cannot possess the characteristic singular behavior over the whole field of observation; therefore, the actual time dependent stress field needs to be computed in order to extract realistic data from experimental records. This is no mean task.

There exists thus at present considerable doubt that in transient crack tip motions and possibly as a result of waves interacting with the running crack tip the present methods of experimental analysis yield always definitive estimates of an instantaneous stress intensity factor. At the same time one must remember that one may be involved in such a detailed description of the crack tip behavior that characterization of fracture by a stress intensity

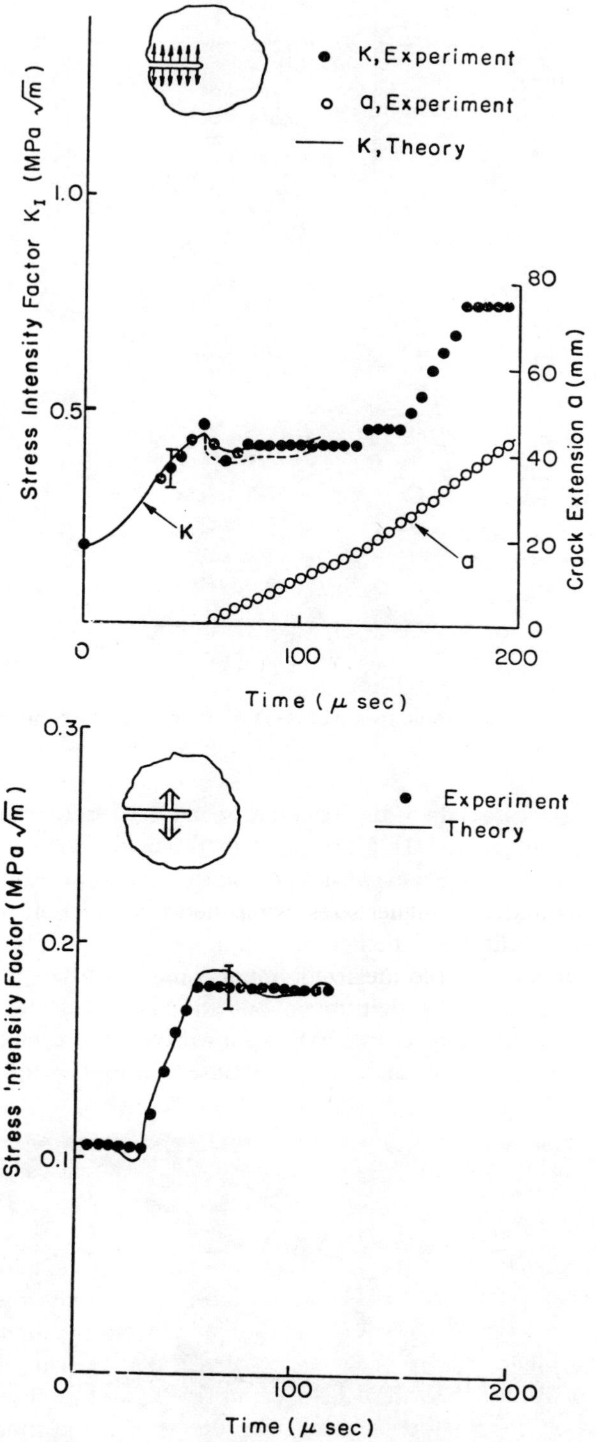

Figure 2. History of stress intensity factor for distributed and local pressure on crack surfaces; comparison of theory and experiments [28].

factor may not be reasonably justified on physical grounds.

Today the major methods for such determinations are the photoeleastic and the caustic methods. In addition to the requirements of thicker plates for fringe sensitivity the photoelastic method must gather information from a larger domain about the crack tip

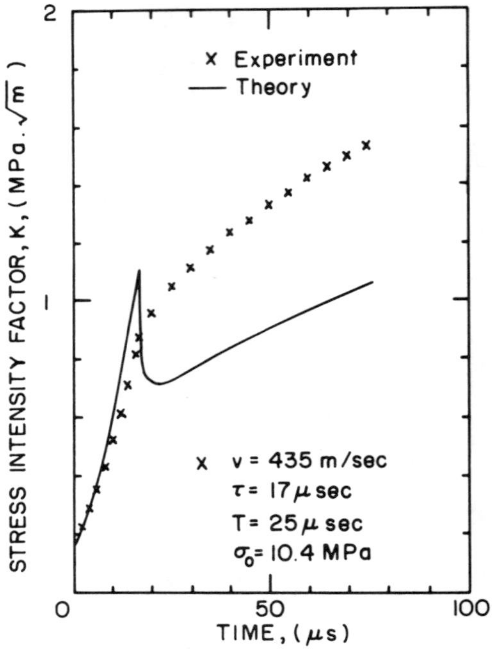

Figure 3. Stress intensity history for uniformly pressurized crack surfaces, but the pressure is approximately twice that in Fig. 2a.

than the caustic method, especially if data evaluation involves more terms in an eigenvalue expansion than the singular one [31]. The caustic method, which does not respond to the "zero order term" in an eigenvalue expansion, gathers information from a smaller domain about the crack tip but becomes inaccurate when the radius of that domain is less than 1/2 of the plate thickness [32].

If the method of caustics is used another complication may arise when micro-fractures at the crack tip generate waves that move out radially from the moving tip. This phenomenon occurs with the tougher materials such as steels and even PMMA [cf. Fig. 4]. The shadow image of these radial waves are superposed on that of the caustic, which is a

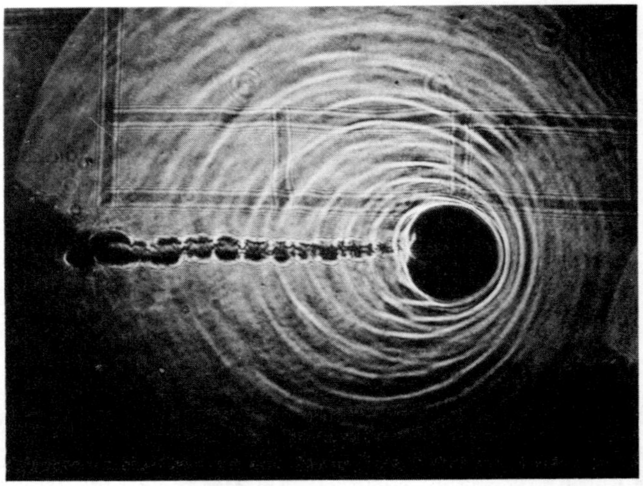

Figure 4. Stress waves from local fracture sources emanating from running crack tip in PMMA.

nearly circular feature centered on the crack tip. Depending on the instant of observation the radial wave may just happen to be slightly inside, on, or outside of the caustic trace and thus modify or partially obscure the size of the caustic.

While this fact requires special care in the use of caustics as a measurement technique which does not arise with the use of photoelasticity, the method of caustics appears to be today the only viable one for stress intensity determination when the solid is stress optically insensitive or too thin for photoelasticity to be useful. On the other hand, one should not deduce from this observation that the photoelastic method yields necessarily more reliable data for stress-birefringent materials. As mentioned before the photoelastic method draws on information from a relatively large domain about the crack tip, especially when "higher order terms" in an eigenvalue expansion are used in data-fitting. Under these circumstances the photoelastic method tends to time-average crack-tip information so that the true time history of the crack-tip deformation field becomes less precise.

3. Non-ideal fracture

With few exceptions real materials do not conform closely to idealized point fracture at the crack tip. Instead, fracture occurs in a zone ahead of the crack tip, as, e.g., for the more ductile solids in quasi-static fracture. It is generally thought that the main feature of deviations from idealized behavior is due to the occurrence of plastic flow at the crack tip, with the simultaneous appearance of voids that grow in the zone of intense plastic

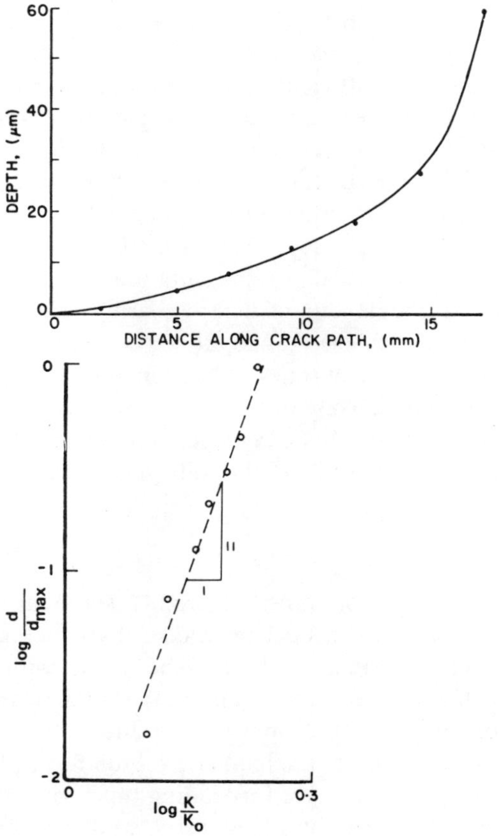

Figure 5. Surface roughness along crack path and as a function of the stress intensity factor [12/II].

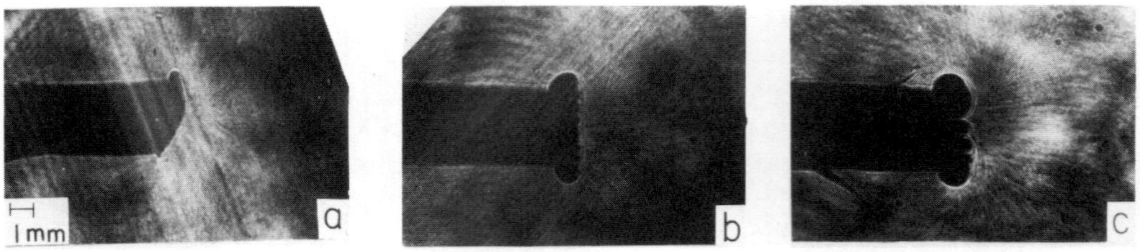

Figure 6. Caustics generated at individual micro-fractures at front of running crack at three different, increasing stress intensity factors [2/III].

deformation. While this widely held viewpoint is no doubt appropriate for ductile solids, we note that even so-called brittle materials do not seem to lead to near-ideal fracture behavior when cracks propagate dynamically.

In experiments conducted on Homalite 100, usually considered as a glass-like, brittle solid in both static and dynamic fracture experiments, one observes a measurable fracture zone even in the apparent absence of plastic deformation. At least three observations speak for this statement, namely, (a) the stress intensity dependent roughness of the fracture surface, (b) direct observation of discrete fracture sources at the crack tip via high-speed photography, and (c) stress waves generated by discrete micro-fractures at the crack tip which travel out radially from the point of fracture as the crack tip passes. The dependence of surface roughness and the size of the fracture zone on the instantaneous stress intensity factor is illustrated in Fig. 5 and has been discussed in [12/II]. The evidence of individual fractures along the crack front at three different levels of crack tip stresses is shown in Fig. 6 as excerpted from [12/II]. Stress waves generated from such micro-fracture in PMMA are illustrated in Fig. 4. One notes that these latter stress waves appear only in connection with sufficiently high crack tip stress when enough energy is released in the micro-fracture to generate a macroscopically observable deformation.

The occurrence of micro-fractures close to the crack tip has, no doubt, an important influence on the time scale of the fracture process. First, we note, that a finite time τ_* is required to establish a process zone either for a virgin crack or for a crack in propagation; for the latter case this would be $\tau_* = \alpha/v$ where α is the size of the process zone and v is the crack speed. Next, one observes that representation of a "singularity field" requires that α be small compared to the domain of the singularity field. For transient situations it may not always be possible to achieve that condition because there is not enough time to develop that field over an area several times as large in extent as α. Thus the details of the fracture process at the crack tip is likely to follow the idealized concepts rather poorly and a better understanding results only if one considers the effects of multiple microfractures there. These observations are derived from the following experiments.

3.1. Initiation of fracture and arrest

A criterion for fracture initiation or transition from a stationary to a moving crack tip would be, in the simplest form of the idealized model that crack growth ensues when the stress intensity factor reaches a critical value. In experiments with plates of Homalite 100 [12/I] one finds that such a simple criterion applies only for relatively low loading rates. Figure 7 shows times of initial crack propagation resulting from loading a semi-infinite crack in an infinite plate with the crack loaded by a uniform pressure which increases roughly linearly with time. One notes that for loading times longer than about 40–60 μsec fracture occurs at a constant value of stress intensity, while for shorter times the required stress intensity rises sharply. A similar behavior has been observed with High Density

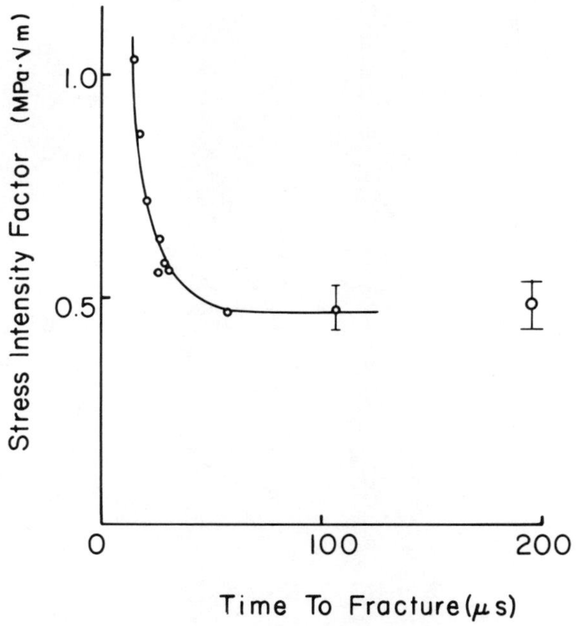

Figure 7. Stress intensity factor at initial propagation of crack under stress wave loading [12/I].

Polyethylene and with Polytetraflourethylene [32] although the time scale is somewhat larger in this case which fact may be due to the difference in viscoelastic behavior of the two materials [34] if viscoelasticity plays any role at all. Also, from tests on steels researchers at the Stanford Research Institute infer similar behavior under short duration stress pulses; i.e. pulses of shorter duration require higher stress levels to initiate fracture [35].

In our experiments [12/I] crack arrest occurs very abruptly, i.e. without a discernable deceleration phase although the time resolution of about 5 μsec may be too coarse to make such a statement; yet, this observation is in agreement with those of Schardin [36] on silicate glass and of Kanninen on steel [18], but is at variance with Bradley and A.S. Kobayashi's experiments [37] as well as the data of Kalthoff et al. [38]. However, in agreement with the latter authors we find that the stress intensity factor at arrest is lower than that for quasi-static initiation by about ten per cent [12/I].

3.2. Relation between crack velocity and stress intensity

Often the effect of stress intensity on crack speed is presented in $K-v$ curves for different materials alluded to in the earlier discussion on idealized fracture, which implies a one-to-one and instantaneous correspondence between these quantities. In experiments with Homalite 100 we find that such a one-to-one correspondence does not appear to exist but rather, that in spite of sizeable changes in stress intensity the crack is reluctant to change speed: the crack speed changes primarily when rapid changes in stress intensity occur due to waves impinging on the moving crack tip [12/III].

In Fig. 8 we show typical histories of crack tip position and the stress intensity factor, determined by the method of caustics as detailed in [28]. Multiple experiments yield the plot shown in Fig. 9, where large variations in stress intensity match up with crack tip velocities that are constant within experimental accuracy.

One notes further from the upper portion of Fig. 9 that a considerable range of velocities may be achieved depending on how rapidly the crack tip is loaded: the crack

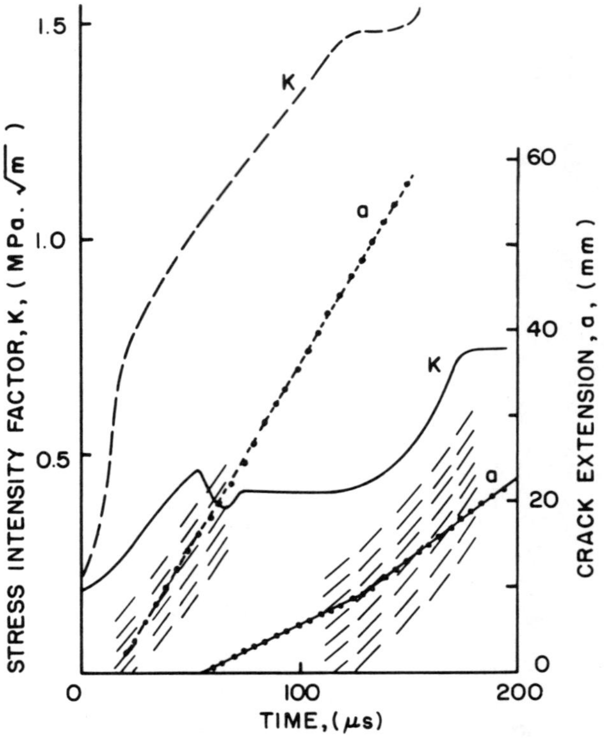

Figure 8. Histories of stress intensity factor and corresponding measured crack extensions. Short lines indicate slopes expected from a $K - v$ curve.

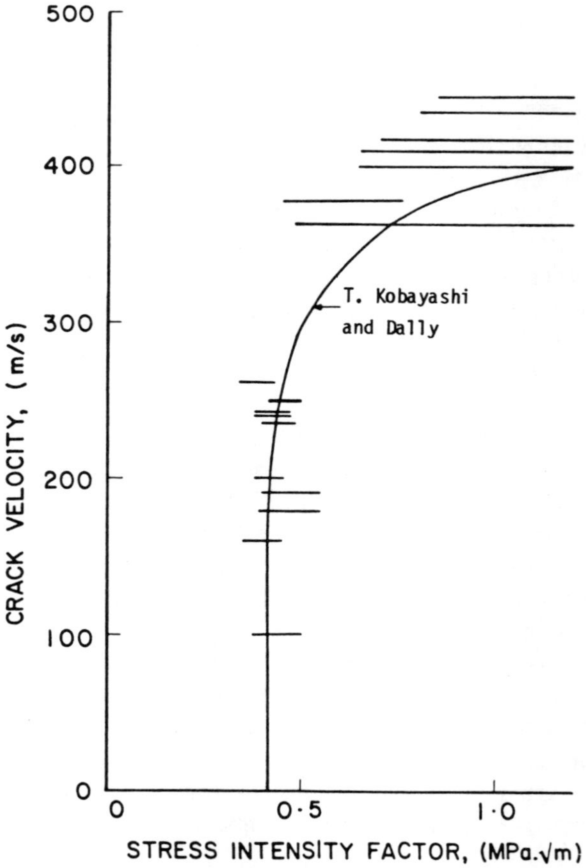

Figure 9. Relation of crack tip velocities in Homalite 100 to stress intensity factors. Suggested $K - v$ curve also shown as solid line [31].

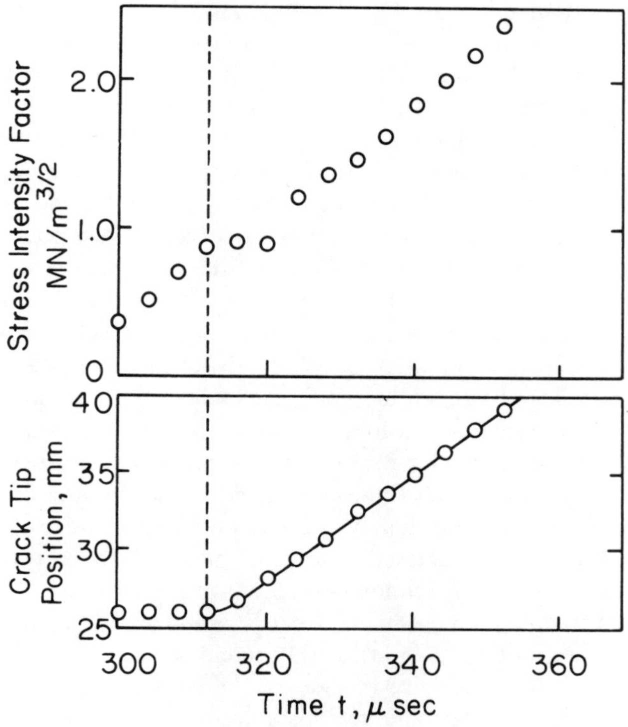

Figure 10. Stress intensity and corresponding crack extension histories for Araldite B [38].

speed seems to be set by the crack tip conditions at initiation and to contradict the existence of a clear upper velocity limit that is sometimes supposed to exist at less than the Rayleigh wave speed. It is unfortunate that virtually all dynamic crack propagation data is reported in the literature in the reduced form of $K - v$ curves without reliable data on the time history of the crack tip position along with indications of measurement precision, nor is there similar data on the corresponding history and precision of the stress intensity factor. There is thus only limited data from sources outside of our own laboratory that impacts this question. Kalthoff [38] has presented data for Araldite B as shown in Fig. 10 where the stress intensity varies by more than a factor of 2 while the crack speed is, within stated accuracy, constant although, according to the corresponding $K - v$ data it should have changed by at least 30 per cent. Similar results may be deduced from the work of *T.* Kobayashi and Metcalf [39] and of Dally and Shukla [40]. Therefore, this phenomenon is probably not totally material-specific. [1]

We are well aware of the problems in spatial resolution in experiments to determine the instantaneous crack tip position and the added difficulties with obtaining the time derivative of such data. These problems are addressed in [31] in this issue in an attempt to deal with the unexpected observation that crack propagation is, to say the least, very insensitive to slow changes in stress intensity [12/III].

In order to further explore this potential effect of measurement precision we refer to Fig. 8: The size of the round symbols represent approximately the range of error in the crack tip position. Also shown in the same figure are slopes of the crack tip history

[1] Recent, as yet unpublished results from tests on PMMA in our laboratory indicate that, within experimental error, the crack speed is virtually constant with varying stress intensity although some speed changes are noticeable, primarily at low stress intensity and crack initiation (crack acceleration).

computed from the solid curve in Fig. 9. We make no claim that this latter curve represents our specific batch of test material. Yet from a practical point of view it is interesting to note the following: The crack propagation rate for the stress history possessing higher magnitudes a major portion of the growth rate falls in the plateau region, while only a very small portion at the beginning might possess a lower crack growth speed. Next concentrate on the other crack growth history, which appears best represented by two straight lines, the change in slope (velocity) being associated with the arrival of a reflected wave at the crack tip. The solid curve in Fig. 9 possesses infinite slope for the stress intensity factor range right after the crack starts to propagate: The corresponding velocities are indeterminate. The higher stress intensities in the same crack growth history generated a continuously increasing velocity which is not matched by the constant value after the event of wave reflections. Note also that there is only a rather small range of velocities and stress intensity factors where the "unique" character of the $K-v$ curve can be studied because a unique relation between K and v exists neither in the vertical stem region (infinite slope) nor in the horizontal plateau region (zero slope).

There is another factor that needs to be considered in this discussion, namely the effect of specimen geometry, or more specifically, the effect of specimen size on the conditions at the crack tip. This question is addressed in more detail in a later section where the effect of multiple reflections from the specimen boundaries are considered.

We note first that to date all data on crack propagation speed excepting those reported in [12/I-IV] have been obtained on relatively small specimens where multiple wave reflections interact with the crack tip. In view of the earlier discussion on the precision with which one is able to determine the instantaneous stress intensity factor in the presence of wave transients one is thus not very certain of the crack tip stress history. Clearly, the effect of test specimen geometry demonstrated by Kalthoff [17,31,38] can be due to stress histories resulting from differences in direct load transmission to the crack tip as well as wave reflections from the boundary. Furthermore, we note that the difference between the $K-v$ curves obtained with small specimens and the data resulting from large plate specimens [12/III] where wave reflections occur in a very discernable way, is significantly larger than the difference derived from small specimens [17]. There appears thus some orderliness in this confusing picture which suggests that multiple interactions of the crack tip with reflected waves affect the crack growth history in an as yet unknown way. This observation would seem to have possibly serious implications for application of laboratory data to large structures undergoing dynamic failure if the observations discussed here apply to other materials.

3.3. Maximal crack speed

While idealized fracture predicts the Rayleigh wave speed to be the limiting speed of crack growth, engineering materials as well as polymeric stimulants exhibit much lower crack growth rates. The reason for the reduced rate may be varied according to what material is involved. Certainly the effect of plasticity, of hole growth and of (non-linearly) viscoelastic behavior are important. However, regardless of these effects the process of generating micro-fractures and their subsequent coalescence at the crack tip requires an interaction between these micro-fractures which, in turn, requires time. It is thus readily understandable that cracks cannot propagate with near wave-speed velocities as long as multiple micro fractures play a role. One would expect thus that when fracture occurs by "fairly clean cleavage" as in a crystal crack speeds should be higher; this view is born out by the data on several crystals reported by Field [41].

A similar result of elevated crack speed should result if it were possible to reduce the cohesion of a material without changing its elasticity (wave speeds): With reduced cohesion the extent of micro fracture at the crack tip would also be reduced so that this source of crack speed reduction would be eliminated or minimized. Recent studies in our laboratories at Caltech have therefore been directed to study crack growth in plates along lines of weakly re-healed cracks. Initial results, using a very thin layer of a low viscosity oil as a "healing agent" indicate crack speeds on the order of 80% of the Rayleigh wave speed where the virgin material sustained crack speeds on the order of only 40% of that speed.

3.4. Crack branching

Crack branching is a phenomenon that has attracted much attention because of its importance in comminution and "porosity generation" in rock formations. On the basis of studies in optically clear Homalite 100 we view branching as a natural outgrowth of the crack propagation process: Direct high-speed photographic observation shows [12/II] that at sufficiently high crack tip stresses associated with branching, the microfractures which are generated away from the main crack axis grow and turn smoothly away from the main crack. These micro-fractures lead to initial branches that do not span the specimen thickness but are clearly of three-dimensional character. The thought that branching is associated with the generation of microcracks has been explored by Congleton [42] who attempted to examine for a quasi-static situation the effect of a small crack in close vicinity to a major crack. The quasi-static solution to that problem for the interacting stress field has been supplied by Pucik [43]. Carlsson discusses micro-fracturing at the crack tip in steels [44] and Dally draws on Congleton's results in discussing branching [45]. In these discussions the details of micro-fractures with regard to size, initial orientation and growth history were rather unspecified.

The real time photography in [12/II] appears to be, to date, the most direct documentation of this micro-fracturing process. Specifically, in view of Fig. 11 from that reference it appears that micro-cracks are generated off axis and parallel to the main crack; their growth occurs first parallel to the main crack and then a smooth turning occurs from that direction in the branch direction(s) much in agreement with the results indicated by Pucik [43,12/III].

Whether a crack branch is successful or not appears to depend on the interaction of

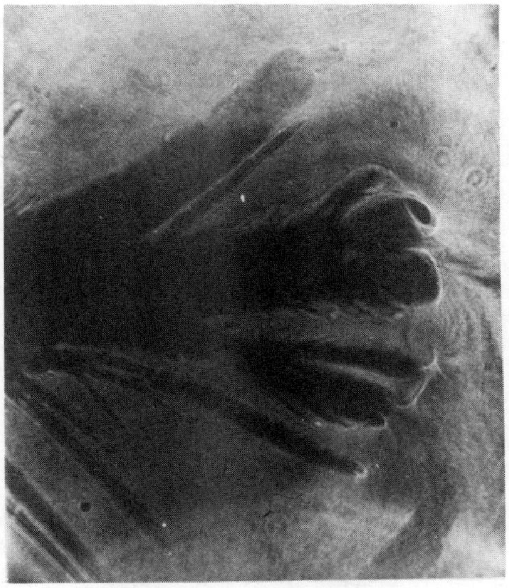

Figure 11. Development of small off-axis cracks into branches by smooth turning of the secondary cracks (field of view approximately 3 mm × 3 mm).

several branches and on the cracks associated with their generation/propagation. A certain statistical quality thus characterizes the branching phenomenon [12/III]. References [12/III], [46], and [47] review branching in some more detail.

Finally we comment on the often held notion that a major difference in dynamic fracture as opposed to static fracture is the crack tip speed: crack tip speed is viewed like an independent or controlling variable. For example branching is often considered to be the result of high velocity, probably as a remnant of Yoffe's argument on incipient branching. In contrast we see in all our past experiments the crack tip stress as the primary source for the physical phenomena that accompany crack growth. Thus surface roughness is directly related to stress intensity [12/II]; the often employed characterization of "mirror, mist, and hackle" is, in reality not distinct but, in terms of surface roughness characterizable in terms of a continuous roughness distribution (cf. Fig. 5). Since roughness is an indication of micro-fracturing, this observation is also reasonable. Furthermore, because microfracturing depends on the stresses at the crack tip, branching should depend primarily on the stress intensity, with velocity only an accompanying feature of the process, at most of modifying influence on the stress field at the crack tip; in this belief we agree with Congleton [42]. It appears to us therefore that the instantaneous stress intensity is, in general, a better parameter for the description of damage at the crack tip, while its history will determine the macroscopic growth rate. If the fracturing medium possesses strongly dissipative properties (viscoplasticity) the picture is more complicated.

4. Effect of stress waves on crack growth

The feature which distinguishes dynamic fracture from quasi-static behavior is the presence of stress waves. These waves arise either due to the stresses released from the crack tip at fracture or due to the externally applied loads. The propagation of these waves is governed by the geometry of the specimen and therefore poses a restriction on the direct application of analog experimental results to full-scale engineering problems unless similarly detailed wave analyses are performed. The frequency content of these stress waves is determined by the loading conditions and the details of the microfracturing processes, while the thickness of the specimen determines whether dispersion occurs as the waves propagate. Since the analytical models are based on two-dimensional elastodynamic theory, comparison of the experimental results with the analytical models will be appropriate only when the experimental conditions preclude wave dispersion. As indicated before, the effect of wave dispersion, especially in specimens of small dimensions, makes the interpretation of dynamic fracture experiments complicated, to say the least.

When stress waves reflected from the specimen boundaries return to the crack tip, they alter the crack tip stress state and this can result in a change of crack speed, or if the stress wave was of high intensity cause branching [12/III,IV]. If the specimen dimensions are small, a number of reflections from the boundary arrive at the crack tip in rapid succession or overlap and, in view of the limited spatial and temporal resolution in measurements, the interpretation of the effect of distinct stress waves may become impossible.

In order to study the progressive loss of wave distinction we conducted experiments on specimens of systematically decreasing size. Figure 12 shows fractured specimens from four experiments in which the loading conditions were identical, but for which specimens the lateral dimensions were decreased progressively. In the two smaller specimens the main crack follows an oscillating crack path, while in the larger specimens, the crack path is straight along the axis of symmetry. The differences in the crack paths are then entirely due to the differences in the stress field caused by the number of stress waves reflecting

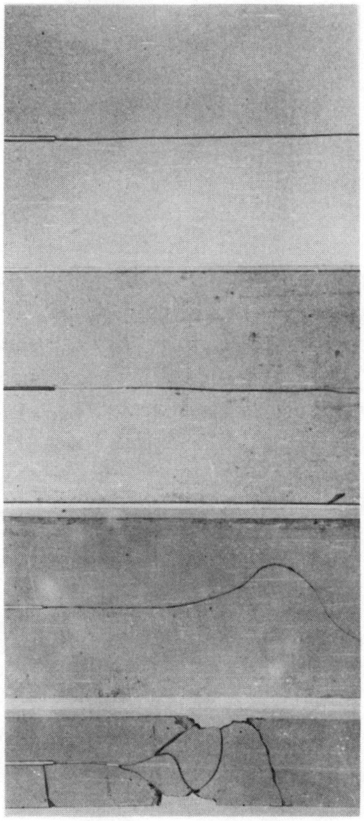

Figure 12. Specimens of varying dimensions moved to crack fractured by identical pressure on crack faces.

from the top and bottom boundaries [12/IV]. In the narrowest specimen, the stress field resembles a time dependent bending field with a large crack-parallel tensile stress, which leads to crack path instability in the sense of Cotterell and Rice [48]. In connection with the earlier discussion on the $K - v$ relation and the observed constancy of the velocity one finds that in these experiments the velocity varies along the curved path between 350 to 450 m/sec, the variation being apparently associated with wave interaction. However, the details of the wave history at the crack tip, which moves at any instant at some angle to the boundary, are not clearly enough established to make a definitive statement about the effect of such an interaction of the crack tip with stress waves.

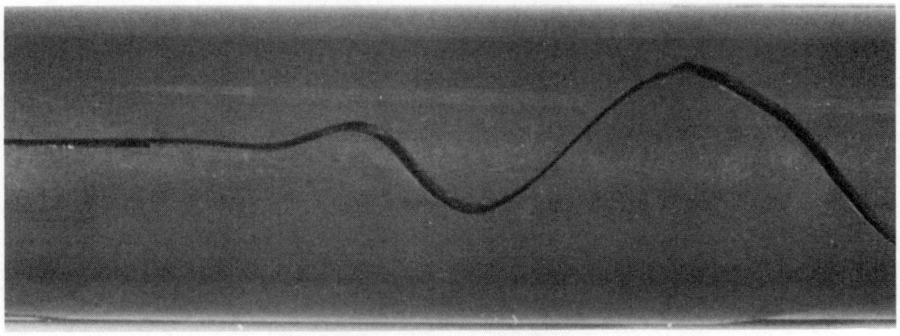

Figure 13. Crack propagation path in PVC pipe.

This oscillating crack path is similar to that observed in pipes under internal pressure (see Fig. 13). [2] Recalling that for a pipe with a longitudinal crack and loaded by internal pressure a tensile stress acts in the crack-parallel direction, it is easy to account for the path instability. The oscillating path is then due to stress waves which shed from the crack tip and propagate in the circumferential direction.

5. Conclusion

We have summarized the differences between idealized and real fracture behavior resulting from stress wave induced loading. A key ingredient in that difference is the development of a zone of micro-fractures at the crack tip. It is difficult to cope with this physical reality in analytical modelling. Yet, also on the experimental side of research there remain questions with regard to definition of crack tip stresses and velocity determination. Some of these questions, primarily those related to the definition of crack tip stress field can be addressed by more careful analytical evaluations while others must rely on the development of more refined measurement techniques.

Acknowledgements

This review and substantial amounts of the work reported here have been supported by a grant from the National Science Foundation (MEA-8215438) and by a past grant by the Office of Naval Research (ONR N00014-78-C-0634). Also, the whole editorial effort for the two issues of the *International Journal of Fracture* on dynamic fracture by the first author was a service for the dynamic fracture community under the National Science Foundation effort.

References

[1] J. Hopkinson, *Original Papers*, Cambridge University Press (1901) 316–320.
[2] B. Hopkinson, *Scientific Papers* (1912) 461.
[3] K.B. Broberg, *Journal of Applied Mechanics* 22 (1955) 317–323.
[4] H. Kolsky and D.G. Christie, *Transactions, Society of Glass Technology* 36T (1952) 88.
[5] D.R. Curran, D.A. Shockey, and L. Seaman, *Journal of Applied Physics* 44 (1973) 4025.
[6] S.N. Zhurkov and V.S. Kuksenko, *International Journal of Fracture* 11 (1975) 629–639.
[7] D.A. Shockey, L. Seaman and D.R. Curran, *International Journal of Fracture* 27 (1985) 145–157.
[8] J.D. Achenbach, in *Mechanics Today* 1. Edited by Nemat-Nasser, Pergamon Press (1974) 1.
[9] L.B. Freund, in *Mechanics Today* 3. Edited by Nemat-Nasser, (1976) 55–90.
[10] C. Atkinson and J. Eshelby, *International Journal of Fracture* 4 (1968) 3–8.
[11] L.B. Freund, *Journal of Elasticity* 2 (1972) 341.
[12] K. Ravi-Chandar and W.G. Knauss, *International Journal of Fracture* I: 25 (1984) 247–262; II: 26 (1984) 65–80; III: 141–154; IV: 193–204.
[13] L.R.F. Rose, *International Journal of Fracture* 12 (1976) 799–813.
[14] J.D. Eshelby, in *Inelastic Behavior of Solids*. Edited by Kanninen et al. McGraw-Hill, New York (1970) 111.
[15] A.A. Wells and D. Post, *Proceedings, SESA* 16 (1958) 69.
[16] W.M. Beebe, Ph.D. Thesis, California Institute of Technology (1966).
[17] J.F. Kalthoff, *International Journal of Fracture* 27 (1985) 277–297.
[18] M.L. Kanninen, *International Journal of Fracture* 27 (1985) 299–312.
[19] E. Yoffe, *Philosophical Magazine* 42 (1951) 739.
[20] A.W. Maue, *Zeitschrift fur Angewandte Mathematik und Mechanik* 34 (1954) 1.
[21] D.D. Ang, Ph.D. Thesis, California Institute of Technology (1958).
[22] A.T. De Hoop, Doctoral Dissertation, Technische Hogeschool, Delft (1958).
[23] K.B. Broberg, *Arkiv. fur Physik* 18 (1960) 159.

[2] We are indebted to R. Pfaff, graduate student at the Graduate Aeronautical Laboratories, Caltech, for providing this photograph from his studies on pipe fracture.

[24] J.W. Craggs, *Journal of Mechanics and Physics of Solids* 8 (1960) 66.
[25] B.R. Baker, *Journal of Applied Mechanics* 29 (1962) 449.
[26] P.B. Crosley and E.J. Ripling, *Journal of Basic Engineering* 91 (1969) 525–534.
[27] L.B. Freund, *Journal of Mechanics and Physics of Solids* 21 (1973) 47.
[28] K. Ravi-Chandar and W.G. Knauss, *International Journal of Fracture* 20 (1982) 209–222.
[29] C.C. Ma and L.B. Freund, Brown University Report (September 1984).
[30] K. Ravi-Chandar, and W.G. Knauss, SM Report 84–18, California Institute of Technology (1984).
[31] J.W. Dally, W.L. Fourney and G.R. Irwin, *International Journal of Fracture* 27 (1985) 159–168.
[32] A.J. Rosakis and K. Ravi-Chandar, Galcit SM Report 84-2, California Institute of Technology (March 1984).
[33] J.G. Williams and J.M. Hodgkinson, *Proceedings, Royal Society* of London A375 (1981) 231–248.
[34] W.G. Knauss, in *Proceedings, Sixth International Conference on Fracture*, S.R. Valluri et al. (editors) New Delhi, India, December 1984, Pergamon Press.
[35] D.A. Schockey, J.F. Kalthoff, H. Homma and D.C. Erlich in *Proceedings, Workshop on Dynamic Fracture*, California Institute of Technology, Pasadena, CA (Feb. 1983) 57–71.
[36] H. Schardin, in *Fracture*. Edited by Averbach et al., Technology Press MIT, John Wiley & Sons and Chapman & Hall (1959).
[37] W.B. Bradley and A.S. Kobayashi, *Engineering Fracture Mechanics* 3 (1971) 317.
[38] J.F. Kalthoff, in *Proceedings, Workshop on Dynamic Fracture*, California Institute of Technology (1983) 11–35.
[39] T. Kobayashi and J.T. Metcalf, in *Crack Arrest Methodology and Application*, ASTM STP 711 (1980) 128.
[40] J.W. Dally and A. Shukla, *Mechanics Research Communication* 6 (1979) 239.
[41] J.E. Field, *Contemporary Physics* 12 (1972) 1.
[42] J. Congleton et al., *Philosophical Magazine* 16 (1967) 749.
[43] T.A. Pucik, Ph.D. Thesis, California Institute of Technology (1972).
[44] J. Carlsson, in *Transactions of the Royal Institute of Technology*, Stockholm; (a) Vol. 189 (1962) 2–55; (b) Vol. 205 (1963) 3–38; (c) Vol. 207 (1963) 3–26.
[45] J.W. Dally, *Experimental Mechanics* 19 (1979) 349.
[46] J.P. Dempsey and P. Burgers, *International Journal of Fracture* 27 (1985) 203–213.
[47] M. Ramulu and A.S. Kobayashi, *International Journal of Fracture* 27 (1985) 187–201.
[48] B. Cotterell and J.R. Rice, *International Journal of Fracture* 16 (1980) 155–169.

Résumé

On décrit les problèmes de caractérisation du champ de contrainte induit autour de l'extrémité de l'entaille par une onde de tension et les conséquences qui en résultent sur le comportement à la propagation d'une fissure. Après avoir posé les hypothèses soujacentes à un modèle idéal de rupture fragile et dynamique, on passe en revue les déviations observées par voie expérimentale à partir d'un tel modèle. Ces déviations sont attribuées au développement de nombreuses microfissures dans la région de l'extrémité de la fissure. Ces microfissures déterminent un processus d'amorçage de fissure, une vitesse de croissance de la fissure et un phénomène d'arborescence. On examine également de manière critique l'assertion selon laquelle il existe une relation unique qui relie la vitesse instantaneée de la fissure au facteur d'intensité d'entaille instantané dans des solides fragiles. La solution à cette question est matériellement gènée par des limitations dans les techniques de mesure habituelles.

International Journal of Fracture 27 (1985) 145–157.

The micro-statistical fracture mechanics approach to dynamic fracture problems

D.A. SHOCKEY, L. SEAMAN and D.R. CURRAN
Poulter Laboratory, SRI International, Menlo Park, CA 94025, USA

(Received September 1, 1984; in revised form November 9, 1984)

Abstract

Continuum fracture mechanics concepts should be applied to solve dynamic fracture problems wherever a continuum approach can provide sufficient answers. Many dynamic fracture problems, however, involve multiple cracks or voids and are not well-suited to the relatively simple continuum approach. This paper describes a statistical fracture mechanics concept on a microscopic scale and illustrates its use for the case of shock-wave-induced ductile spall fracture. The paper further shows how micro-statistical fracture mechanics (MSFM) merges with continuum fracture mechanics by treating a macrocrack propagating in a DCB specimen using MSFM and the MSFM parameters deduced from the spall work.

Thus, the two approaches are consistent. Although significantly more complicated, the MSFM approach promises to be helpful in improving our understanding of the influences of microstructure on toughness and in extending continuum approaches to more ductile materials and to smaller specimens.

1. Introduction

Problems of dynamic fracture are commonly discussed in terms of single cracks embedded or moving in continuous bodies. However, many dynamic fracture problems are not single-crack problems and cannot be treated readily by a continuum approach. For example, impact erosion of piping and valves in coal gasifiers, explosive excavation of rock in tunnels and mines, and penetration of armor plate by projectiles are problems involving many simultaneously active cracks as well as other simultaneously occurring failure processes. The situation is illustrated in Fig. 1, which shows a polished and etched cross section through the impact site of a steel plate that was impacted at 6 km/s by a 6-mm-diameter water-filled polycarbonate sphere. Failure processes in evidence are ductile fracture (voids beneath the crater), brittle fracture (cracks extending inward from the crater profile), adiabatic shear banding (the white etching bands extending inward from the crater profile and also those connecting the ductile voids beneath the crater), the α (bcc) to ϵ (hcp) to α (bcc) phase transformations indicated by the hemispherical dark-etching region encompassing the crater as well as the homogeneous plastic flow indicated by the curvature in the originally straight and parallel rolling texture. All five failure processes occur more or less simultaneously, and each process influences the other as it occurs.

The complexity of the situations precludes a treatment by classical fracture mechanics concepts. A micro-statistical fracture mechanics approach is more appropriate in which the development of individual microfailure features is treated statistically, and the effect of this development on the state of stress or strain is accounted for continuously.

The first purpose of this paper is to describe this MSFM approach and illustrate its

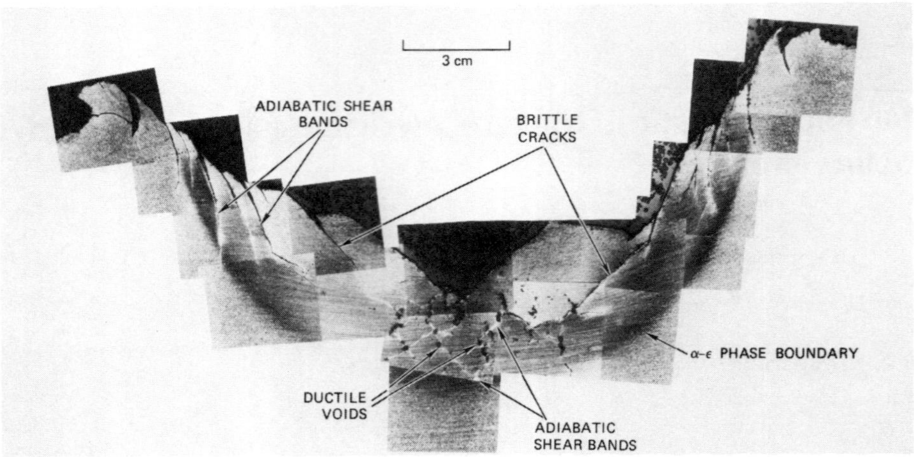

Figure 1. Polished and etched cross section through the impact site of a steel plate that was impacted at 6 km/s by a 6-mm-diameter water-filled polycarbonate sphere.

application using as the example the case of shock-induced ductile voids in steel [1]. A similar microstatistical treatment can be applied to shear bands [2]. The second purpose is to show that the micro-statistical approach merges with the continuum approach. This is done by using the void kinetics equations and the data generated in the shock wave experiments to compute the behavior of a macrocrack propagating in a DCB specimen [3].

2. Shock-wave-induced ductile fracture

Dynamic ductile fracture can be studied without the complications caused by other accompanying failure processes in simple, well-defined, plate impact experiments. The

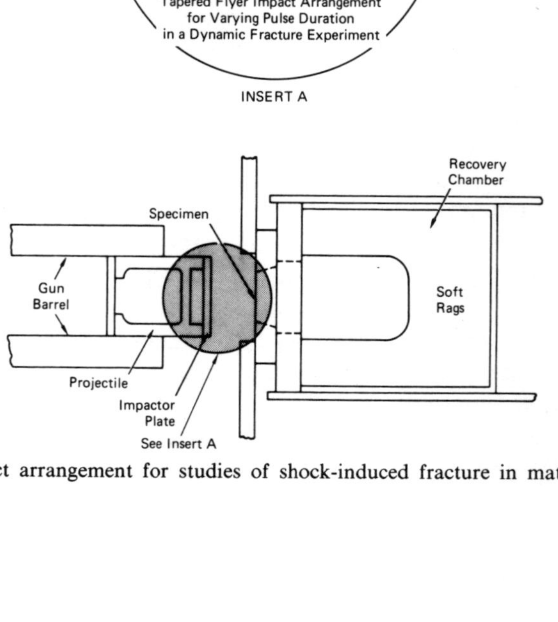

Figure 2. Plate impact arrangement for studies of shock-induced fracture in materials under uniaxial strain conditions.

Table 1. Elemental composition of XAR30 armor steel

Element	Wt%
C	0.28
Mn	0.94
P	0.010
S	0.025
Si	0.66
Cr	0.59
Mo	0.21
Zr	0.14
B	0.0014

Table 2. Tensile properties of XAR30 armor steel *

Orientation	0.2% yield strength (ksi)	Tensile strength (ksi)	Elongation (%)	Reduction in area (%)
Longitudinal	208	262	14.5	49.8
Long transverse	216	253	10.5	39.8
Short transverse *	211	261	10.0	12.5

* Short transverse properties are the average of three tests performed at SRI. All other values are the average of two tests by Hickey [4].

arrangement shown in Fig. 2 is commonly used to produce a short-lived tensile pulse in a specimen under a one-dimensional-strain state. Here a flyer plate mounted on a cylindrical projectile is accelerated in a gun barrel by sudden release of pressurized gas and is made to impact a specimen plate at the gun muzzle. Care is taken to ensure simultaneous impact between the two plates at all locations, so that lateral strains are zero until unloading waves from the periphery can propagate into the specimen. The specimen plate is decelerated in a chamber filled with energy-absorbing material to prevent subsequent and uncontrolled impact loads.

Table 3. Dynamic fracture experiments on XAR30 armor steel

Experiment number	Specimen thickness (mm)	Flyer thickness (mm)	Angle of taper (deg)	Flyer velocity (mm/μs)	Remarks
2024-1 **	7.62	3.81	5.7	0.451	Full fracture; clear gage record
2024-2 **	10.1	4.44	5.7	0.357	Large continuous fracture; clear gage record
2024-3	5.08	2.24	5.7	0.226	Numerous microfractures
2024-4	12.1	5.58	6.84	0.200	Numerous microfractures
2024-5	2.54	1.12	2.3	0.259	Numerous microfractures

** Instrumented with back-surface ytterbium piezoresistive stress transducers.

The initial compression waves produced in the colliding plates on impact reflect at the plate boundaries to produce in the specimen a tensile pulse whose amplitude is specified and controlled by the impact velocity. The pulse duration is controlled by the plate thicknesses. Thus, if the flyer plate is tapered, as shown in the inset in Fig. 2, the duration of the pulse experienced by the specimen varies with location, and the dependence of ductile fracture damage on time-at-stress can be investigated in a single experiment. If a series of experiments are performed at different impact velocities, the effect of stress can also be studied. The procedure is illustrated here for an armor steel.

Plate impact experiments

High-hardness armor steel made by Great Lakes Steel Company and designated as XAR30 was selected for investigation. Chemical composition and tensile properties are

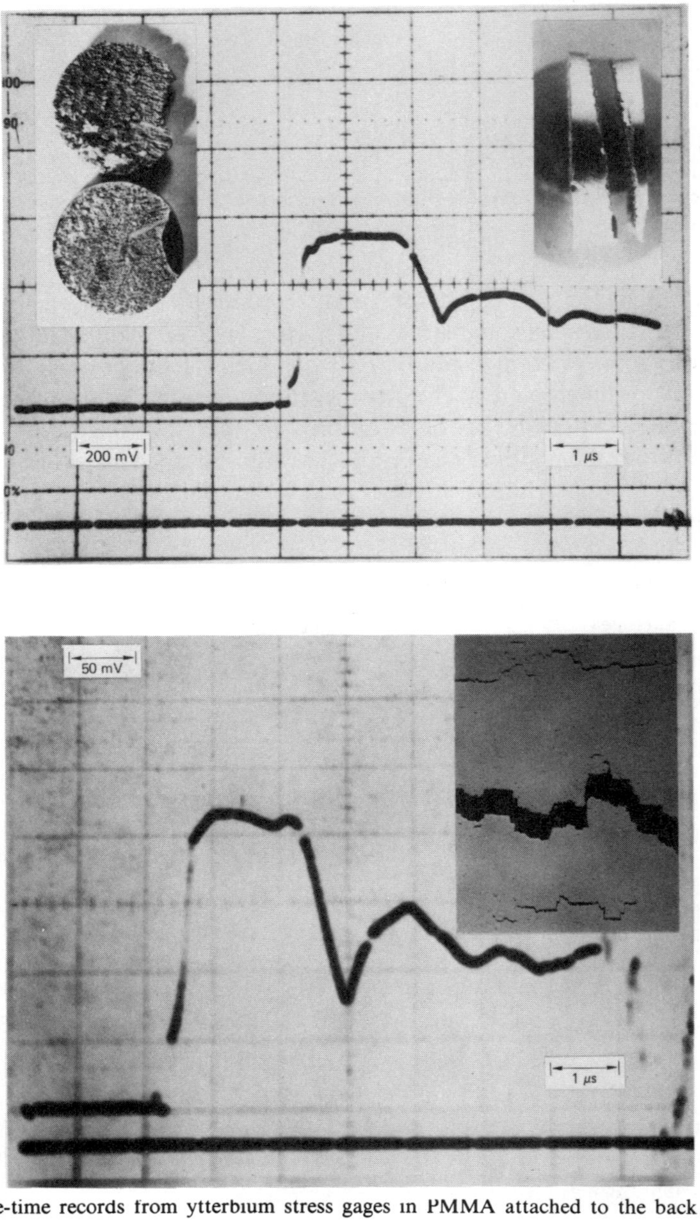

Figure 3. Voltage-time records from ytterbium stress gages in PMMA attached to the back surfaces of plate impact specimens.

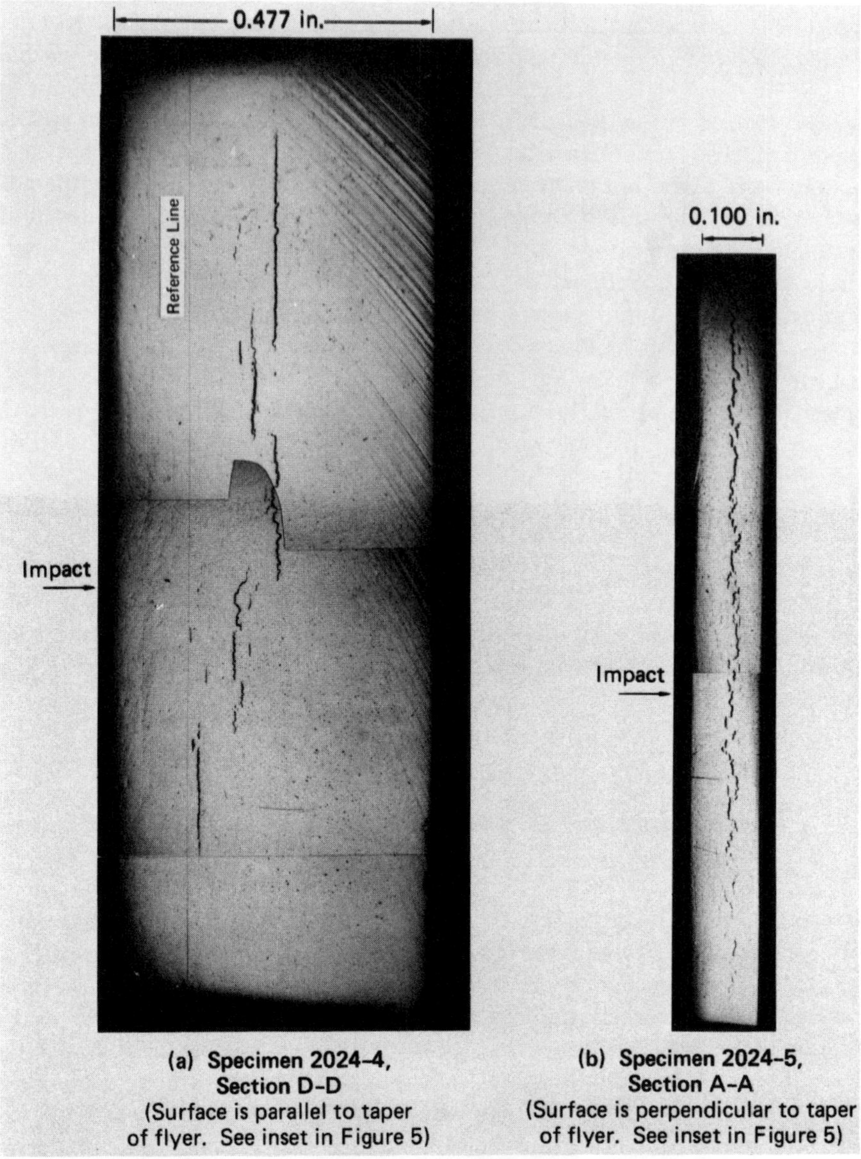

Figure 4. Polished cross sections showing internal shock-induced fracture damage.

given in Tables 1 and 2. Specimen disks 3.81 cm in diameter were machined to various thicknesses (see Table 3) and mounted in the target plate. The edges were beveled at an 8-degree angle to allow the specimen to release easily from the target plate on impact and fly into the catcher tank.

Tapered flyer plates 5.72 cm in diameter and having an average thickness half that of the specimen were machined from the same material and mounted on the front of a 15-cm-long aluminum projectile. The angle of taper varied for different experiments from 2.3 to 6.8 degrees, as shown in Table 3. The impacting surfaces were ground flat and parallel to within 0.01 mm.

The projectile was accelerated through the barrel of the gas gun by sudden release of compressed helium, and careful alignment of the flyer plate and specimen resulted in flat plate impact. The impact velocity was measured by electrical contacts at the gun muzzle.

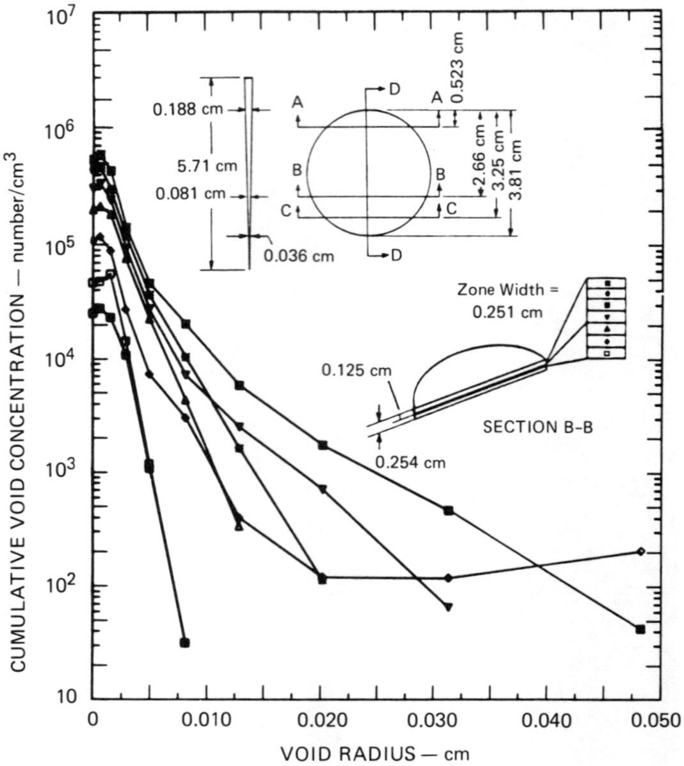

Figure 5. Measured void size distributions on section B-B of specimen 2024-5.

Wave profiles were recorded close to the rear surface of two of the specimens. Ytterbium piezoresistive stress transducers, mounted in 9.5-mm-thick blocks of C-7 epoxy that were glued to the specimens, produced the oscilloscope traces shown in Fig. 3. Both records show the Hugoniot elastic limit (HEL), a flat-topped loading wave, and a clear fracture signal. The HEL measures the yield strength of the material under dynamic uniaxial strain conditions and is useful in specifying the constitutive equation of the armor. The flat-topped loading wave indicates that impact planarity was good and that significant attenuation did not occur before fracture. The second peak in the records, known as the fracture signal, is caused by a reloading recompression of the transducers by waves emanating from the internal fracture surfaces as they form and grow.

Recovered specimens were sectioned to reveal the internal fracture damage, as shown in Fig. 4. Specimens 2024-3 and 2024-4 were sectioned to produce a cross section in the direction of maximum taper, and Specimen 2024-5 was sectioned parallel to the direction of taper at three locations as indicated in the inset in Fig. 5.

The fracture damage in Specimens 2024-3, 2024-4, and 2024-5 was analyzed quantitatively by counting and measuring the traces of the microfractures on the polished section surfaces. Measurements were made by using a large area record reader with which an operator positions a cross hair on one end of a void trace, pushes a button to record the coordinates, and repeats the process for the other end. A simple computer program uses these data to compute the length, orientation, and position of the trace within the specimen. The size distributions on the sectioned surfaces are then converted to actual void size distributions per unit volume by means of a statistical transformation implemented in the BABSI computer code. [5]

The results of this quantitative damage analysis are void size distribution curves for various positions on a cross section. An example is shown in Fig. 5.

Analysis of the data

Equations governing nucleation and growth of voids have been deduced from previous plate impact studies [5,6] to be of the form

$$\dot{N} = \dot{N}_0 \exp \frac{\sigma - \sigma_{n0}}{\sigma_1} \tag{1}$$

and

$$\dot{R} = \frac{\sigma - \sigma_{g0}}{4\eta} R \tag{2}$$

where $\dot{N}$ and $\dot{R}$ are time-rate-of-change of the number and size of the voids, respectively; σ is the instantaneous value of the tensile stress; and the remaining symbols are empirically determined, material-specific parameters. These equations have successfully described dynamic tensile damage in a wide variety of materials and hence appear to be quite general.

Values for the fracture parameters for XAR30 steel were obtained from the observed damage in the three tapered flyer experiments in the following way. The average void nucleation rate (total number of voids divided by the nominal duration of the tensile stress) was plotted versus peak stress in tension to obtain a first estimate of the nucleation threshold stress (σ_{n0}) and the other nucleation parameters (σ_1 and $\dot{N}_0$). The shape parameter $R = R_1$ of the observed distribution was plotted versus tensile impulse (peak tension times the duration of the tension) to determine the nucleation size (R_0) and the growth rate parameter, η.

After the initial estimates of all five parameters were determined from plots, one-dimensional trial calculations were performed to approximate the impact conditions at several points in the target. These calculations were repeated with different parameters in (1) and (2) until the computed and measured damage compared satisfactorily. Then a two-dimensional calculation was performed to simulate the entire impact. It was not necessary to modify the parameters further and repeat the two-dimensional simulation. The fracture parameters found for XAR30 armor are listed in Table 4. For comparison, the parameters for Armco iron and aluminum, obtained in earlier work [6], are also included in the table.

Figure 6 indicates the agreement obtained between measured and computed damage. This figure shows the void-size distribution on the planes of maximum damage on two sections of Specimen 2024-5. The crack size distributions are exponential in the computa-

Table 4. Dynamic fracture parameters for XAR30 armor steel, Armco iron, and 1145 aluminum

Parameter	Units	XAR30	Armco Iron	1145 Al
η, material viscosity	dyn-s/cm^2	4545	417	75
σ_{g0}, growth threshold	dyn/cm^2	-1.0×10^8	-2.0×10^8	-4.0×10^9
R_0, nucleation size	cm	4.0×10^{-3}	5.0×10^{-5}	1.0×10^{-4}
$\dot{N}_0$, threshold nucleation rate	no./cm^3/s	4.0×10^8	4.6×10^{12}	3.0×10^9
σ_{n0}, nucleation threshold	dyn/cm^2	-2.5×10^{10}	-3.0×10^9	-3.0×10^9
σ_1, nucleation sensitivity	dyn/cm^2	-1.786×10^9	-4.56×10^9	-4.0×10^8

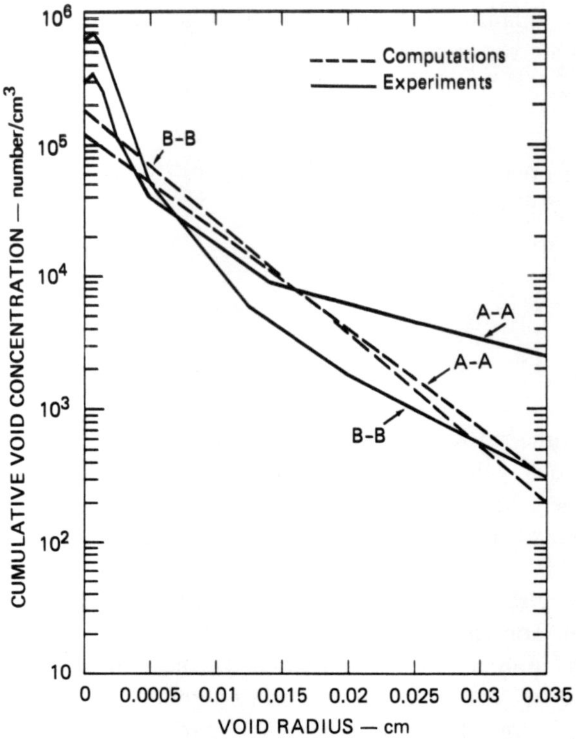

Figure 6. Comparison of measured and computed damage in XAR30 armor steel on planes of maximum damage in tapered flyer impact experiment 2024-5.

tions and therefore appear as straight lines, thus, the detail of the curves is not predicted. However, average numbers and sizes of voids are predicted satisfactorily.

Summary of progress in MSFM and required research efforts

Although we may not have captured the details of the void nucleation and growth processes, we believe that a working treatment exists for computing the development of ductile fracture damage. For example, (1) and (2) and the dynamic fracture parameters of Table 4 can be used to compute the void population and the resulting stress relaxation beneath the crater in Fig. 1. The benefits of this approach include the capability to establish threshold conditions for fracture damage, a framework for predicting residual strength, and an avenue for linking microstructure to dynamic fracture response. However, this treatment is empirical and cannot yet be extrapolated with confidence to other materials and loading conditions. More work is required to merge theoretical nucleation and growth models with observed void behavior.

Even less well understood than nucleation and growth is void interaction and coalescence. A reliable description of this final stage of failure is required for the computation of propagation of macroscopic cracks, full spall, and fragmentation. Some insight into the coalescence process for the XAR30 armor steel under plate impact loading conditions is provided in Fig. 7, which shows two parallel but nonplanar fractures in the process of coalescing. A profusion of tiny microfractures has formed in a path linking the tips, suggesting that coalescence is a nucleation and growth process on a smaller scale. Advances in our understanding and our capability to quantitatively characterize the coalescence process can be expected in the near term through efforts to quantitatively

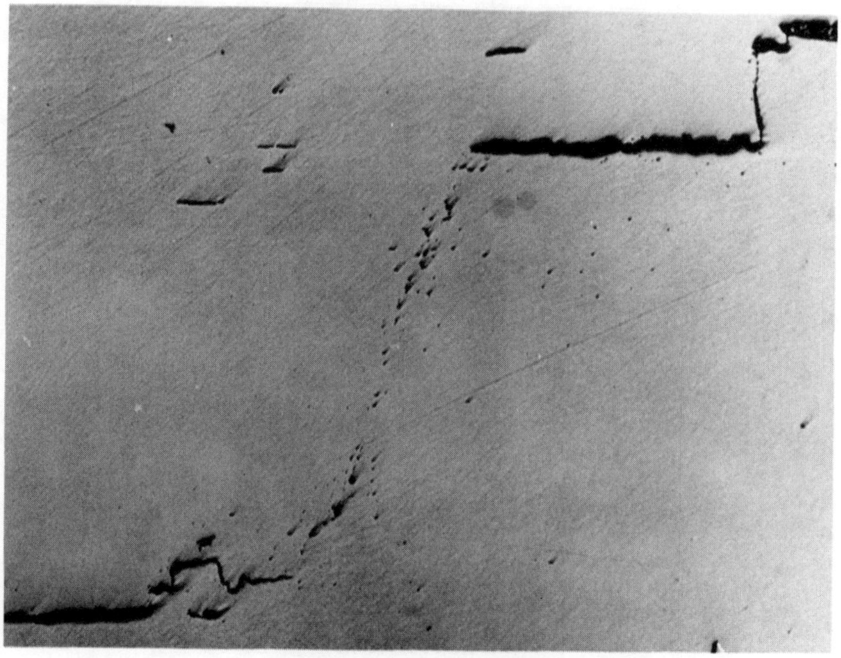

Figure 7. Polished cross section in XAR30 armor steel showing incipient coalescence of two planar macrofractures.

evaluate fracture surface features and by application of a topographic analysis technique being developed by Kobayashi [8]. Because the distribution of dimples on a fracture surface indicates the distribution of the void population, an estimate of the final void size distribution is obtained by counting and measuring the dimples [7]. Estimates of the void size distribution at various stages before fracture may be obtainable by Kobayashi's technique of determining fracture surface topology and superimposing to various degrees the topographs of conjugate fracture surfaces. Thus, advances in our ability to model void coalescence can be expected in the near future.

3. Microcrack propagation

The micro-statistical fracture mechanics (MSFM) described above must merge with the continuum fracture mechanics approaches discussed in the other papers of these two issues. The classical fracture mechanics approach treats macrocrack propagation as the movement through a homogeneous medium of a continuous crack front separating broken from unbroken material. In reality, however, macrocrack propagation occurs by nucleation, growth, and coalescence of voids in the process zone slightly ahead of the crack front. For further details on this topic related to brittle materials see the paper by W.G. Knauss and K. Ravi-Chandar in this issue. Figure 8, for example, shows microvoids that have formed at the tip of a notch in a Charpy specimen. A comparison of the pre-notch-tip damage with the fracture damage produced by plate impact shows that the damage morphology is the same and suggests strongly that the fracture phenomenology is also the same, namely, nucleation, growth, and coalescence of individual microvoids. Further evidence for crack growth by pre-crack-tip void activity can be found on fracture surfaces, which exhibit the characteristic dimples, the remnants of severed voids.

To determine whether the microfracture nucleation and growth functions extracted from plate impact experiments could be used to describe microfracture activity ahead of a

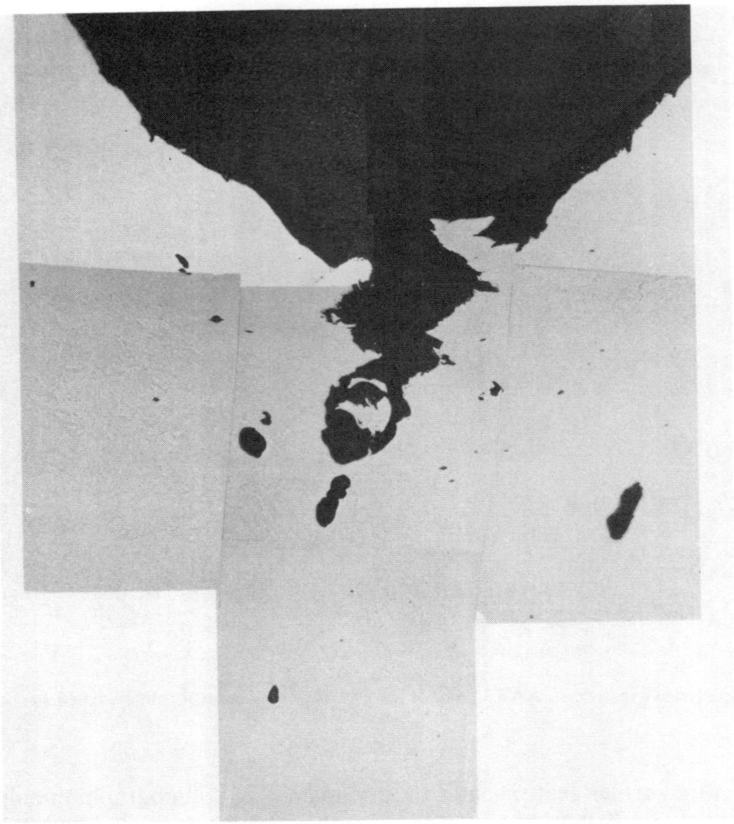

Figure 8. Polished midplane section of a Charpy impact specimen of A533B steel that was loaded to a subcritical deflection showing the ductile crack emerging from the notch base and isolated voids near the crack tip.

propagating crack, we performed a computational simulation of a double-cantilever-beam (DCB) crack arrest experiment using the MSFM approach and the dynamic fracture data from the plate impact experiments described in the previous section.

A two-dimensional wave propagation code was used to compute the stress histories experienced by the material at the moving crack tip in a DCB crack arrest specimen of 4340 steel quenched and tempered to a strength level of about 200 ksi. The tests were performed by workers at the Battelle Memorial Institute. The experiment setup is shown in Fig. 9. A wedge attached to the crosshead of a tensile machine was forced downward between pins passing through holes at the notched end of the specimen until a crack emerged from the notch tip and propagated into the specimen. The crack opening displacement at instability, the crack velocity, and propagation distance were measured. Reference [9] gives more details.

The computation was accomplished in two stages. The first stage covered elastic deformation: the measured wedge opening displacements and forces were applied to the finite element grid to obtain the elastic stress distribution in the specimen at the point of crack instability. The method of dynamic relaxation [10] was used to accomplish this economically. When the static, elastic solution was obtained to within about 1%, yielding was allowed to occur.

The second calculational stage began with instability of the blunted notch and continued until many microseconds after arrest. During this stage, the true dynamic stress field was computed and allowed to dictate the stress state and fracture activity at the propagating crack tip. Void nucleation and growth were computed using the model of (1)

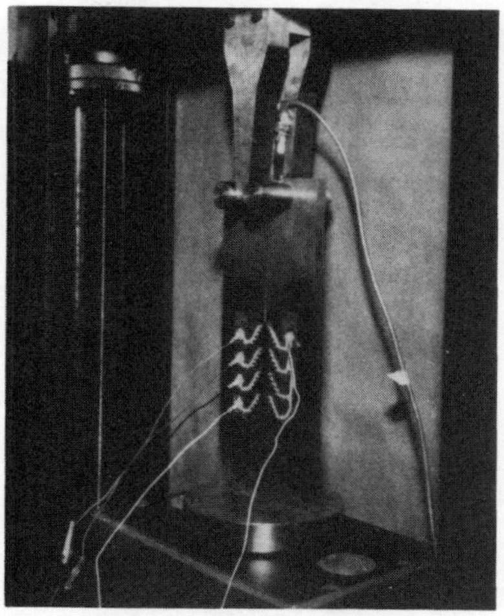

(a) WEDGE LOADING ARRANGEMENT FOR DCB SPECIMEN SHOWING THE VELOCITY MEASURING CONDUCTING STRIPS

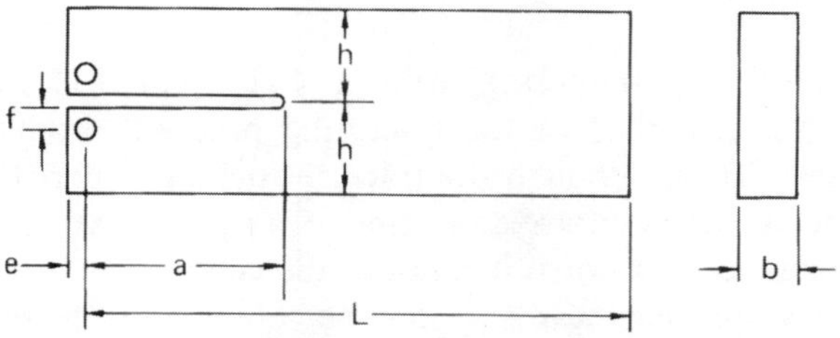

a_o = 2.670 in. (67.8 mm) f = 0.80 in. (20.3 mm)
b = 0.500 in. (12.7 mm) h = 2.500 in. (63.5 mm)
e = 0.63 in. (16.0 mm) L = 11.36 in. (305.0 mm)
Pin Diameter 1.00 in. (25.4 mm)

(b) DIMENSIONS OF TEST PIECES

Figure 9. Crack arrest experiment on a wedge-loaded DCB specimen of 4340 steel performed at Battelle Memorial Institute.

and (2) and parameters from Table 4 that were deduced from the plate impact work. The strength of the computational cell was gradually reduced with increasing void volume. When void volume equaled cell volume, coalescence was said to be complete, cell rupture occurred, and the crack advanced once cell length. Reference [3] gives details of the computational procedures.

The crack ran three times farther in experiments than calculated and propagated only half as fast. We believe better agreement would result if a finer computational mesh were used; nevertheless, the work shows the feasibility of computing macrocrack behavior from events occurring on a microscopic scale at the crack tip.

An example of the insight into the fracture process made possible by the MSFM approach is given in Fig. 10, which shows the variation with time of the elastic strain energy and the energies absorbed by plastic flow and by microcracking. As the crack tip approaches and loads an elemental cell, the elastic strain energy rises sharply. Soon afterward, plastic flow occurs and the crack tip material begins to absorb the elastic

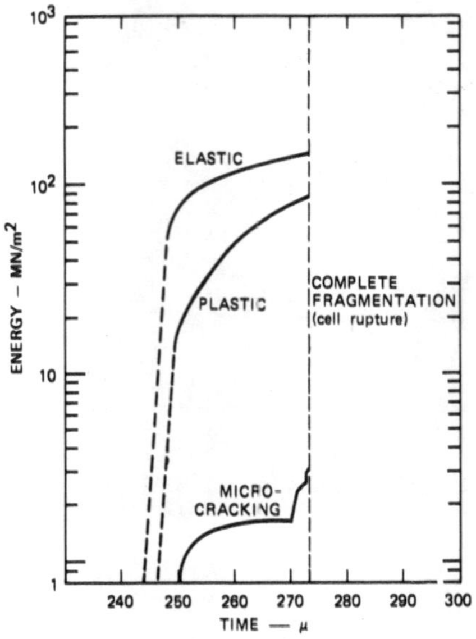

Figure 10. Energy partitioning in a computational cell in a DCB specimen as the cell is traversed by the crack.

energy. A short time later, microfracturing begins and after about 25 μs the cell ruptures. Relatively little energy is absorbed in the fracturing process; about 20 times more is absorbed in plastic flow. Thus, although the fracture toughness derives largely from the plastic work, the microfracturing controls the amount of plastic work because microfracturing terminates the plastic work when it ruptures the cell.

The dynamic propagation toughness K_{ID} may be computed from Fig. 10 by summing the absorbed energies. The resulting value $K_{ID} = 61$ MN $m^{-3/2}$ compares favorably with reported values [9] for static initiation toughness and arrest toughness of 54 MN $m^{-3/2}$ and 68 MN $m^{-3/2}$, respectively. This K_{ID} value should be considered as a rough estimate because of the coarseness of the computational grid used and because of the approximate nature of the model.

The motivations for using more detailed and physically realistic models to compute cract propagation and arrest are threefold. First, such approaches promise to provide the link between continuum mechanics parameters and microscopic material behavior. For example, it would be a great advance to be able to derive the fracture toughness from a knowledge of individual contributing factors on the microstructural level, such as plastically dissipated energy and the energy consumed in the microscopic fracture process itself. Second, the more detailed approach used here is not restricted by geometry, strain rate, or even stress state. Therefore, the results of detailed calculations may be helpful in extending continuum approaches, which have the advantage of simplicity, to apply to more ductile materials and to smaller specimens. Third, the role played by microstructure in controlling toughness can be studied and perhaps understood in a more quantitative way.

Thus, the MSFM merges with continuum treatments much in the way that quantum mechanics merges with Newtonian mechanics. Both have their advantages and disadvantages and special areas of applicability. Continuum treatments have the advantage of simplicity and low cost and therefore should be used whenever they can provide sufficient answers. However, the more complex and expensive MSFM treatment may be the only viable approach when detailed information is needed regarding size, velocity, and trajectory distributions of fragments ejected from a plate of armor steel after impact, as exemplified in Fig. 1.

Acknowledgments

This paper was prepared under contract DAAG29-81-K-0123 with the US Army Research Office. The technical work was supported by the Army Materials and Mechanics Research Center, Watertown, MA, under Contract No. DAAG46-72-C-0182 and by the Electric Power Research Institute, Palo Alto, CA, under Contract No. RP 499-1. A similar account was originally published in *Material Behavior Under High Stress and Ultrahigh Loading Rates,* 29th Sagamore Army Materials Research Conference, Plenum Press, New York (1983).

References

[1] L. Seaman, and D.A. Shockey, *Models for Ductile and Brittle Fracture for Two-Dimensional Wave Propagation Calculations,* Final Technical Report AMMRC CTR 75-2, Army Materials and Mechanics Research Center, Watertown, Massachusetts (1975).

[2] L. Seaman, D.R. Curran and D.A. Shockey, *Material Behavior Under High Stress and Ultrahigh Loading Rates,* 29th Sagamore Army Materials Research Conference, Plenum Press, New York (1983) 295.

[3] D.A. Shockey, M. Austin, L. Seaman and D.R. Curran, *Computation of Crack Propagation and Arrest by Simulating Microfracturing at the Crack Tip,* Final Technical Report to Electric Power Research Institute, Palo Alto, California, Contract RP499-1-1 (1976).

[4] C.F. Hickey, Jr., *Toughness Data for Monolithic High-Hardness Armor Steel,* AMMRC FTR 72-3, Army Materials and Mechanics Research Center, Watertown, Massachusetts (July 1972).

[5] L. Seaman, D.R. Curran and R.C. Crewdson, *Journal of Applied Physics* 49 (1978) 5221.

[6] T.W. Barbee, Jr., L. Seaman, R. Crewdson and D.R. Curran, *Journal of Materials* 7 (1972) 393–401.

[7] D.A. Shockey, L. Seaman, K.C. Dao and D.R. Curran, *Journal of Pressure Vessel Technology* 102 (1980) 14–21.

[8] T. Kobayashi and D.A. Shockey, in *Synergism of Microstructure, Mechanisms, and Mechanics in Fracture,* AIME Annual Meeting, Los Angeles, CA, Feb. 27-29, 1984, to appear.

[9] G.T. Hahn, R.G. Hoagland, M.F. Kanninen and A.R. Rosenfield, *Dynamic Crack Propagation,* G.C. Sih, Ed., Noordhoff International Publishing, Leyden (1972) 649.

[10] J.R.H. Otter, A.C. Cassell and R.E. Hobbs, in *Proceedings of the Institute of Civil Engineering* Paper No. 6986 (1967).

Résumé

Les concepts de fracture mécanique d'un continuum devraient être appliqués à la solution des problèmes de fracture de rupture dynamique dès lors qu'une approche par continuum peut fournir des réponses suffisantes. Cependant, plusieurs problèmes de rupture dynamique comportent des fissures ou des cavités multiples et ne sont pas bien adaptés à une approche par continuum relativement simple. Le mémoire décrit un concept de mécanique de la rupture statistique à une échelle microscopique et illustre son utilisation au cas d'une rupture par décollement ductile sous l'effet d'une onde de choc. On montre en outre qu'une mécanique de rupture micro-statistique peut être envisagée dans le mécanisme de rupture d'un continuum en traitant la propagation d'une macro-fissure dans une éprouvette Cantilever et en utilisant ce concept et les paramètres qui y sont liés, déduits du travail relatif à l'écaillage.

Ainsi, les deux approches se révèlent applicables. Bien que considérablement compliquées, l'approche de mécanique de rupture micro-statistique se révèle prometteuse pour améliorer notre compréhension des influences de la microstructure sur la ténacité et pour une extension de l'approche du continuum de milieu continu à des matériaux plus ductiles ou à des éprouvettes plus petites.

International Journal of Fracture 27 (1985) 159–168.

On the uniqueness of the stress intensity factor – crack velocity relationship

J.W. DALLY, W.L. FOURNEY and G.R. IRWIN

Photomechanics Laboratory, Department of Mechanical Engineering, University of Maryland, College Park, MD 20742, USA

(Received September 1, 1984; in revised form November 9, 1984)

Abstract

Experimental results due to several different investigators on the $K_I - \dot{a}$ relationship are reviewed and the apparent differences in results leading to questions regarding the uniqueness of this relationship are discussed. The influence of the errors due to the three dimensional state of stress at the crack tip, the effects of non-singular stresses, velocity, transient loading and velocity measurement is presented. These errors have obscured resolution of the uniqueness question and an experiment is described to resolve the issue.

1. Introduction

The analysis of rapid fracture is of considerable importance in several practical applications including crack arrest in engineering structures, fracture plane control in construction blasting and fragmentation in mining and projectile design. Today one assumes that progressive fracture is driven by the stress field in the local neighborhood of the crack front. A relationship between the crack speed and the local stress field is necessary for any type of dynamic analysis. In homogeneous and isotropic solids which are brittle, non-linear deformations and crack front irregularities are usually limited sufficiently so that the local stress field which is driving the crack at a given velocity can be expressed as a one-parameter representation in terms of the stress intensity factor K_I. The relationship between the stress distribution and crack velocity required for dynamic analysis is, in this instance, a relationship between the crack velocity and K_I.

A more involved relationship is required when the local region near the crack tip exhibits non-linear behavior or if crack front irregularities are not small relative to the size of the stress field that is dominated by the singular term. In any case, the relation between the stress field and the crack velocity used in any dynamic analysis must be established by experimental observation. This paper will review experimental observations of crack speed as a function of K_I for materials in which this relationship, with certain exceptions, might be expected to be a material property.

2. Review of $K_I - \dot{a}$ results

Experiments conducted by the authors over the past decade utilizing dynamic photoelasticity and high speed multiflash recording of propagating cracks in Homalite 100 have provided data on the relationship between the instantaneous stress intensity factor $K_I(t)$ and the velocity of the crack tip. These experiments [1,2,3,4] have been conducted with six

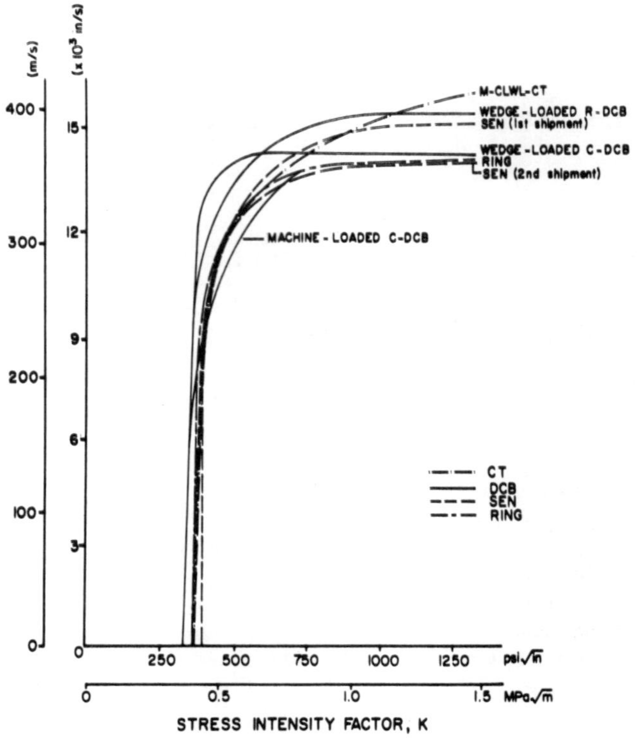

Figure 1. The relationship between crack velocity and the instantaneous stress intensity factor for Homalite 100.

different specimen types and the resulting $K_I - \dot{a}$ curves are illustrated in Fig. 1. Since these results were obtained with two different shipments of Homalite 100, over a three year period, by four different investigators and with experimental errors ranging from ± 10 percent on individual values of $K_I(t)$ to about ± 3 to 5 percent on average values, the agreement was considered satisfactory and certain conclusions were drawn regarding the uniqueness of the $K_I - \dot{a}$ relationship. It was stated [5] that the vertical stem of the $K_I - \dot{a}$ curve and the arrest toughness K_{Im} as determined by photoelasticity are independent of specimen size, configuration and loading system. The horizontal portion of the $K_I - \dot{a}$ curve may not be uniquely related to K_I as there is considerable scatter in the data. In addition, there is significant experimental evidence which indicates that for high velocities the fracture process zone and the branching toughness K_{Ib} may depend upon terms higher than merely the singular portion of the stress field.

Several other researchers including J.F. Kalthoff, K. Ravi-Chandar, W.G. Knauss, A.J. Rosakis, J. Duffy and L.B. Freund have questioned the uniqueness and existence of the $K_I - \dot{a}$ relationship (see Introductory Comments by W.G. Knauss [6]).

Kalthoff [7], utilizing the method of caustics and Araldite B obtained similar $K_I - \dot{a}$ curves; however his results presented in Fig. 2 show a dependence on specimen type with the double cantilever beam DCB-specimen indicating K_I values up to 20 percent higher than those obtained with the single edge notch SEN-specimen. Kalthoff and the authors used static loads applied to the specimen prior to the initiation of the cracks.

Ravi-Chandar and Knauss [8] also employed the method of caustics with Homalite 100. The loading of the crack in an unbounded medium was dynamic and achieved with an electromagnetic loading device for the duration (150 μs) of the experiment. The results of these experiments presented in Fig. 3 indicate the lack of a one-to-one relationship between K_I and $\dot{a}$. Instead, the velocity remains constant as K_I changes and the velocity

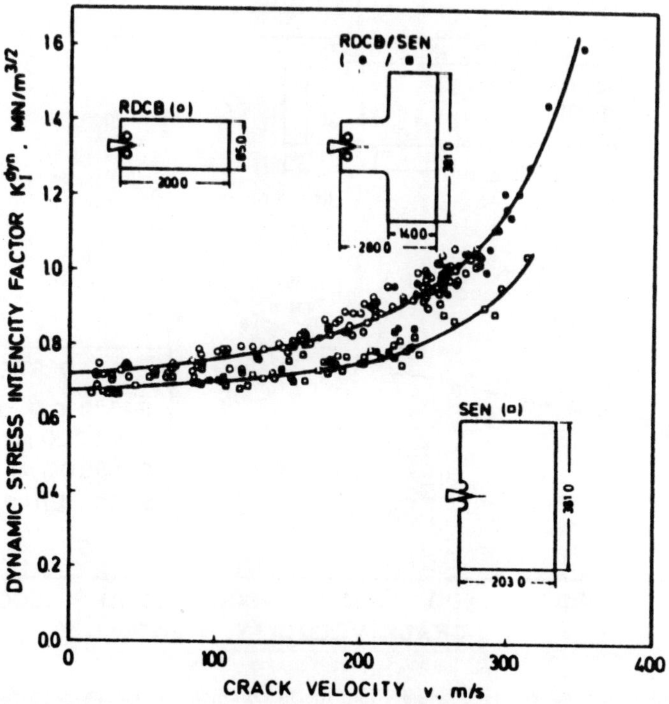

Figure 2. The relationship between crack velocity and the instantaneous stress intensity factor for Araldite B (after J.F. Kalthoff.)

of propagation depended on the value of K_I at crack initiation (K_Q).

Rosakis, Duffy and Freund [9] used reflected caustics to study fracture behavior in 4340 steel. The results obtained shown in Fig. 4 indicate a $K_I - \dot{a}$ relation similar to that obtained by *T.* Kobayashi and Dally [10] utilizing birefringent coatings on 4340 steel. Rosakis et al., concluded that solids fracturing in a ductile manner seem to be well described by a unique $K_I - \dot{a}$ relation while non-unique $K - \dot{a}$ behavior has been noted in brittle polymers and high strength HFX-760 steel.

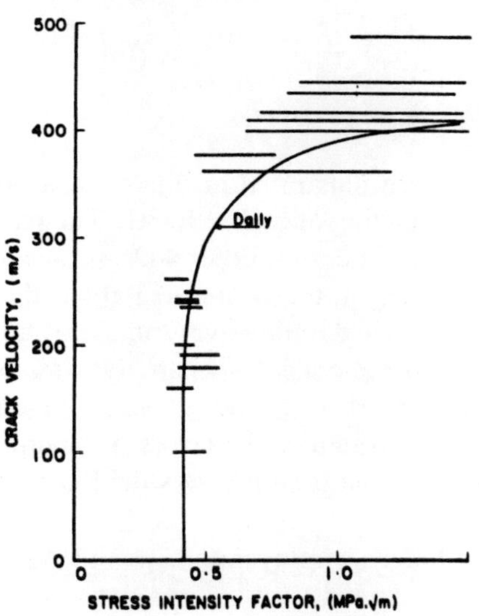

Figure 3. The relationship between crack velocity and the instantaneous stress intensity factor for Homalite 100 (after Ravi Chandar and Knauss.)

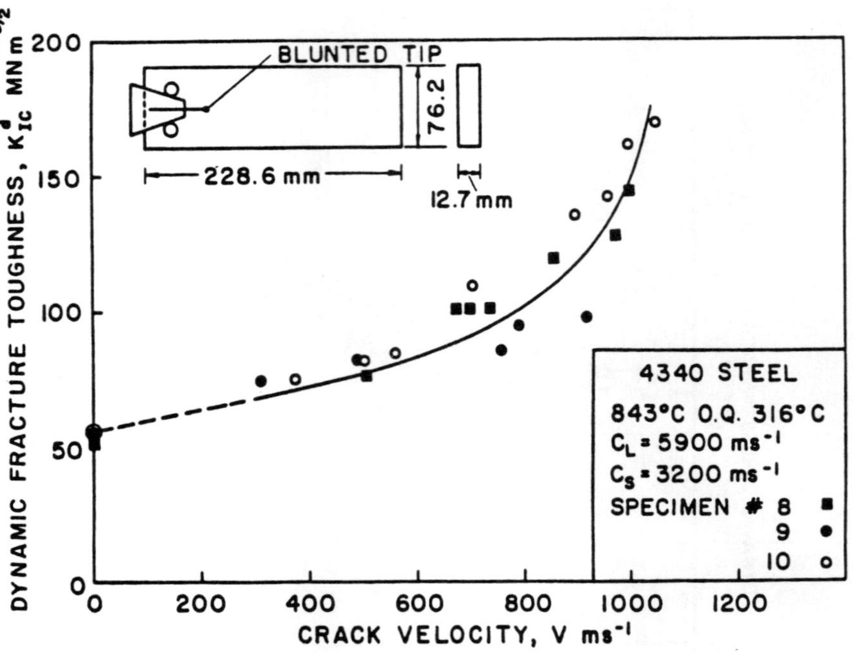

Figure 4. The relationship between crack velocity and the instantaneous stress intensity factor for 4340 steel (after Rosakis, Duffy and Freund).

The difference in the viewpoint on the uniqueness of the $K_\mathrm{I} - \dot{a}$ relation is clear from this brief review. To better elucidate the physical reality, it is imperative that one carefully consider the details of the experiments and the methods used to determine the values of both K_I and $\dot{a}$ from either the isochromatic fringe loops or the shadow spots. Recall also, that the methods of data analysis were being refined concurrently with publication of the results of these dynamic experiments.

3. Experimental methods

A. Photoelasticity

The dynamic isochromatic pattern illustrated in Fig. 5 illustrates the characteristic fringe loops at the tip of a crack propagating at high velocity. The pattern recorded with a spark light source with an exposure time of about 400 ns accurately [11] represents the maximum in plane shear stresses τ_m in the entire field about the crack tip. The method for extracting K_I from the isochromatic data involves the use of the general dynamic solution [12,13] for the state of stress near the crack tip where the stresses are expressed in a power series containing $r^{-1/2}$, r^0, $r^{1/2}$, r, $r^{3/2}$ and r^2 terms. Studies [14,15] show that the higher order terms are necessary to reduce the errors in determining K_I.

The dynamic higher order solution for the stress field is given by [12,13]

$$\sigma_x = \frac{1+\lambda_2^2}{4\lambda_1\lambda_2 - \left(1+\lambda_2^2\right)^2}\left\{\left(1+2\lambda_1^2-\lambda_2^2\right)\mathrm{Re}\,Z_1 - \frac{4\lambda_1\lambda_2}{1+\lambda_2^2}\mathrm{Re}\,Z_2\right\} + \frac{1}{\lambda_1^2-\lambda_2^2}\left\{\left(1+2\lambda_1^2-\lambda_2^2\right)\mathrm{Re}\,Y_1 - \left(1+\lambda_2^2\right)\mathrm{Re}\,Y_2\right\} \tag{1}$$

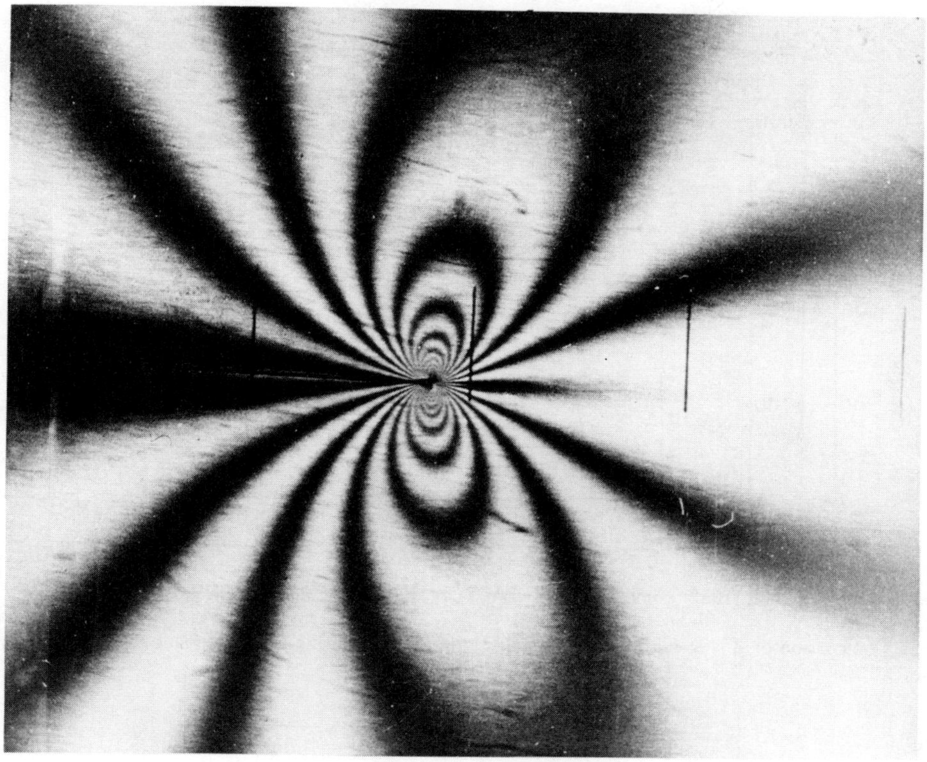

Figure 5. Dynamic isochromatic fringe pattern recorded in Homalite 100 in a light field polariscope.

$$\sigma_y = \frac{1+\lambda_2^2}{4\lambda_1\lambda_2 - (1+\lambda_2^2)^2}\left\{-(1+\lambda_1^2)\ \mathrm{Re}\ Z_1 + \frac{4\lambda_1\lambda_2}{1+\lambda_2^2}\mathrm{Re}\ Z_2\right\}$$

$$+\frac{1+\lambda_2^2}{\lambda_1^2-\lambda_2^2}\{\mathrm{Re}\ Y_2 - \mathrm{Re}\ Y_1\} \tag{2}$$

$$\tau_{xy} = \frac{2\lambda_1(1+\lambda_2^2)}{4\lambda_1\lambda_2 - (1+\lambda_2^2)^2}\{\mathrm{Im}\ Z_2 - \mathrm{Im}\ Z_1\}$$

$$+\frac{1}{\lambda_1^2-\lambda_2^2}\left\{\frac{(1+\lambda_2^2)^2}{2\lambda_2}\mathrm{Im}\ Y_2 - 2\lambda_1\ \mathrm{Im}\ Y_1\right\} \tag{3}$$

where

$$Z_j = Z(z_j) = \sum_{n=0}^{n=N} C_n z_j^{n-1/2}, \quad j=1,2 \tag{4}$$

$$Y_j = Y(z_j) = \sum_{m=0}^{m=M} D_m z_j^m, \quad j=1,2 \tag{5}$$

$$z_j = x + \mathrm{i}\lambda_j y, \quad j=1,2 \tag{6}$$

$$\lambda_j = 1 - (\dot{a}/c_j)^2, \quad j=1,2 \tag{7}$$

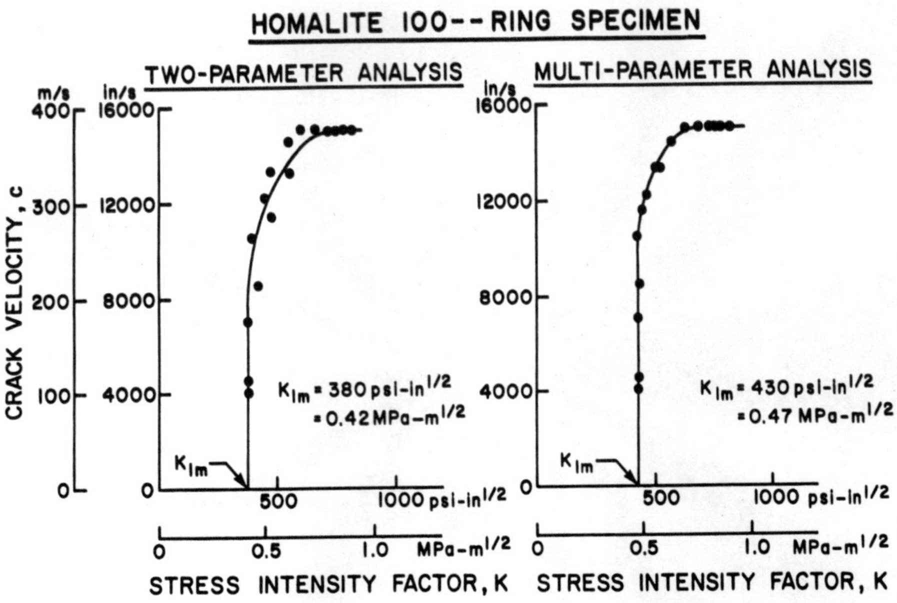

Figure 6. Comparison of $K-a$ curves for a Homalite ring specimen using two-parameter and multi-parameter analyses.

where

$\dot{a}$ is the crack velocity

c_1 is the plate wave speed (plane stress for thin plate analysis)

c_2 is the shear wave speed.

These stresses are used with the stress optic relationship

$$(2\tau_m)^2 = \left(\frac{Nf_\sigma}{h}\right)^2 = (\sigma_x - \sigma_y)^2 + (2\tau_{xy})^2 \tag{8}$$

which yields a system of non-linear equations to be solved for the unknown coefficients C_n and D_m by the overdeterministic method of Sanford and Dally [16]. This method utilizes many data points (40 to 60) where N, r and θ are experimental quantities taken from the field of data provided in the isochromatic pattern. The use of redundant data points permit a least squares analysis which reduces the effect of errors in determining r and θ for specific values of N.

An example of the importance of higher order analysis is demonstrated in Fig. 6 where $K-\dot{a}$ results are given for the same data set using a two parameter analysis and a multi-parameter analysis. Clearly the scatter in the experimental results is greatly reduced with the multi-parameter analysis and the arrest toughness K_{Im} is increased by 13 per cent.

B. Caustics

The caustic pattern illustrated in Fig. 7 was photographed in Araldite B by Kalthoff. The method [17] generally followed in extracting K_I from the shadow spot is to measure the diameter D and relate this single determination to K_I by the formula

$$K_\mathrm{I} = \frac{2\sqrt{2\pi}\,F}{(3)(3.17)z_0 c d_{\mathrm{eff}}} D^{5/2} \tag{9}$$

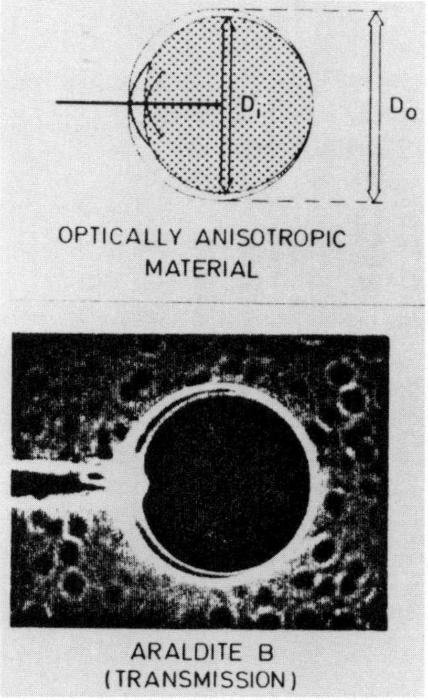

Figure 7. Caustic generated by a crack propagating in Araldite B (after Kalthoff).

where F is a velocity correction factor which can be approximated by

F = $1 - 0.01(\dot{a}/c_R)$ for $\dot{a}/c_R \leqslant 0.4$

c = an optical constant for the material

z_0 = position of shadow image plane from object plane

d_{eff} = effective thickness of the plate.

The general approach followed by investigators reducing the data from caustics is to neglect the effects of higher order terms. This is justified in part since Kalthoff [17] has shown that in optically isotropic materials the r^0 term does not influence the size or shape of the shadow spot and the r term does not affect the diameter D. However, Phillips and Sanford [18] have shown that all higher order terms affect the size and shape of both parts of the double caustics in optically anisotropic materials such as Homalite 100 and Araldite B. The r^0 and r terms produce self cancelling errors when the average of the transverse diameters is used to compute K_I according to (9). The errors produced by the r, $r^{3/2}$ and r^2 terms can be significant and must be taken into account to improve the accuracy of K_I determination from shadow spots.

Another source of error which can affect the accuracy of results from both the method of photoelasticity and caustics is the three dimensional effects in the local neighborhood of the crack tip. Rosakis and Ravi-Chandar [19] and Freund and Yang [20] have recently investigated the three dimensional effect and have found that plane stress conditions prevail at distances from the crack tip greater than half the specimen thickness while no significant plane strain region was detected. These findings were demonstrated in a series of caustic experiments where r_0 the radius of the initial curve was varied from 0.01 to 2 times h_0. The results shown in Fig. 8 indicate very large errors (up to 50 percent) when r_0 is less than $r/h = 0.5$. The results also indicate relatively large scatter even for values of $r_0/h > 0.5$. Considering the fact that r_0 changes with K_I during a dynamic experiment according to

$$r_0 = \left[\frac{3}{2}\frac{K_\text{I}}{\sqrt{2\pi}} z_0 c d_{\text{eff}} F^{-1}\right]^{2/5} \tag{10}$$

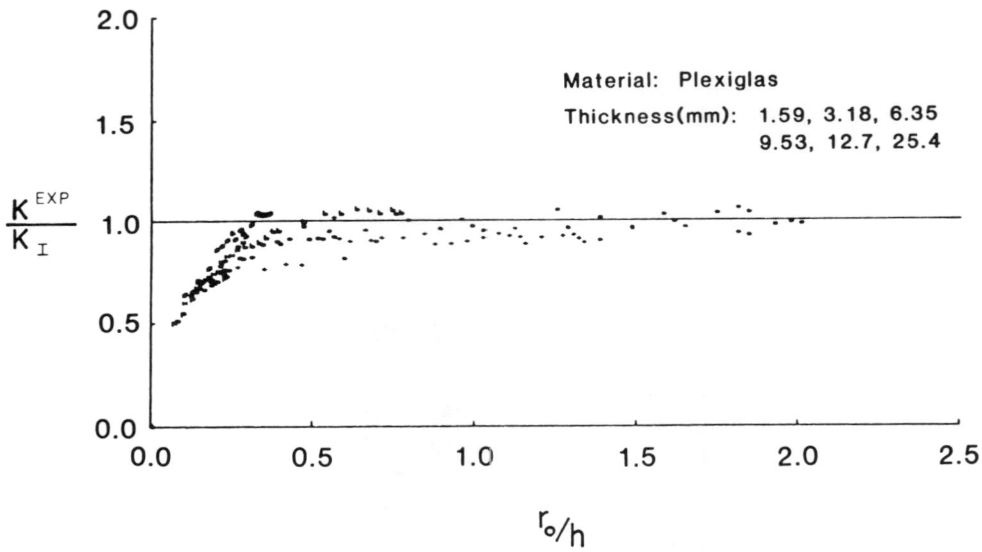

Figure 8. K^{exp}/K_I as a function of r_0/h (after Rosakis and Ravi-Chandar).

it is possible that $r_0/h < 0.5$ in some of the experiments which yielded the results shown in Figs. 1–4.

C. Dynamic effects

Freund [21] has considered stress histories in a plane model containing a crack which is suddenly initiated. Two different problems were considered. In the first, the crack faces were subjected to equal but opposite point loads a distance L from the crack tip and the material is at rest prior to initiation. Freund shows that the stress state near the crack tip is influenced by stress waves and that the magnitude of the computed stress was always less than that for the singular stress component. In as much as Freund did not employ an expansion of the stress field in powers of "r" this fact implies that higher order terms (non-singular) are necessary to account for the more complex stress fields which result from dynamic effects.

In the second problem, Freund considered the same model with the crack faces subjected to a spatially uniform pressure which is applied suddenly at $t = 0$. The solution to this problem shows that the time for the stress state to achieve a singular distribution is surprisingly large (For Homalite 100 70 μs are required before the singular term accounts for 95 percent of the stresses). These results indicate again the importance of higher order terms in the data analysis method to accommodate the non-singular distributions.

The theory and procedures are in place for "higher order analysis" of isochromatic fringe order to accommodate non-singular stress distributions produced by either transient loading or stress wave loading. However, this is not the case with caustics where single measurements of D are made to infer K_I. A more refined higher order dynamic theory must be developed which will utilize multi-point data around the caustic curve to permit the determination of the unknown coefficients arising in a high order representation of the state of stress near the crack tip.

D. Velocity measurements

The velocity measurement is usually made by constructing a graph of the crack length as a function of the time (i.e. "a" versus t). The measurement of crack length on isochromatic photographs is slightly obscured by small circular caustics which appear at the crack tip. With photographs of the caustics the crack tip is severely obscured by the large shadow spot and the exact position of the crack tip is obtained [22] by measuring the front edge of the shadow spot and subtracting kD where k depends upon the material properties. As higher terms can have a significant influence on the horizontal diameter this procedure can introduce error in crack length determinations.

The accuracy of the time measurement is strongly dependent on the instrumentation employed. With high speed photodiodes and digital oscilloscopes it is possible to establish the frame time to ± 0.5 μs. However, with an analog oscilloscope errors of ± 5μs may occur.

Additional error is introduced when the $a - t$ curve is differentiated to obtain the crack tip velocity. The magnitude of this error is dependent on the number of data points available and the accuracies in determining the individual values of a and t for each frame. Considering the methods used to determine a, errors in excess of 10 percent can be anticipated for cracks exhibiting small velocity changes and even larger errors for cracks which are exhibiting more rapid fluctuations in velocity.

4. Conclusions

A review of the experimental results showing the $K_{\mathrm{I}} - \dot{a}$ relation indicated differences which obscure interpretation of the uniqueness of this method of characterizing the control of crack tip velocity. The differences are not large in comparison with the errors which can result in the individual determinations of K_{I} and $\dot{a}$ using either the method of caustics or photoelasticity.

Errors due to the three dimensional state of stresses at the crack tip can be extremely large for both experimental approaches. Errors due to the presence of non-singular terms due to either stress waves or the transient behavior of the state of stress are possibly significant. A procedure to account for these non-singular terms has been developed for photoelastic data; however, a comparable procedure has not yet been developed for caustics. The need for a high order representation is particularly important when transient loading methods are employed as non singular terms are large for a surprisingly long period of time after the application of the dynamic load. Finally, the method of measuring the crack velocity must be markedly improved. It is possible that velocity measurements could be improved by utilizing methods introduced many years ago by Döll [23] or by Kerkhof [24].

More experiments will be necessary to firmly establish the uniqueness of the $K_{\mathrm{I}} - \dot{a}$ curve. It is suggested that experiments similar to those conducted by Kalthoff [7], where several different specimens each progressively smaller are fabricated from the same sheet of material, should be performed using advanced photoelastic procedure for data analysis which avoid the errors noted above. *

The authors believe these experiments will verify the uniqueness of the near-vertical stem of the $K_{\mathrm{I}} - \dot{a}$ curve and will assist our understanding of situations where the non-singular terms affect the fracture process zone. In these regions it will be necessary to use one or more parameters in addition to K_{I} to account for the high velocity behaviors.

Acknowledgements

The authors appreciate the comments of R.J. Sanford, D.B. Barker and R. Chona in discussions leading to this paper.

* The reader is referred to related experiments on successively smaller specimens discussed in [25]. (Footnote added by Editor.)

References

[1] G.R. Irwin et al., "A Photoelastic Study of the Dynamic Fracture Behavior of Homalite 100," U.S. NRC Report NUREG-75-107, University of Maryland (1975).
[2] G.R. Irwin et al., "A Photoelastic Characterization of Dynamic Fracture," U.S. NRC Report NUREG-0072, University of Maryland (1976).
[3] G.R. Irwin et al., "Photoelastic Study of Crack Propagation and Arrest," U.S. NRC Report NUREG/CR-0342, University of Maryland (1977).
[4] G.R. Irwin et al., "Photoelastic Studies of Crack Propagation and Arrest in Polymers and 4340 Steel," U.S. NRC Report NUREG/CR-0542, University of Maryland (1978).
[5] J.W. Dally, *Experimental Mechanics* 19, No. 10 (1979) 349–361.
[6] W.G. Knauss, "Introductory Comments", Workshop on Dynamic Fracture, California Institute of Technology (1–7 February, 1983).
[7] J.F. Kalthoff, "On Some Current Problems in Experimental Fracture Dynamics", Workshop on Dynamic Fracture, California Instiute of Technology (February 1983) 11–25.
[8] K. Ravi-Chandar and W.G. Knauss, "Processes Controlling the Dynamic Fracture of Brittle Solids", Workshop on Dynamic Fracture, California Institute of Technology (February 1983) 119–128.
[9] A.J. Rosakis, J. Duffy and L.B. Freund, "Dynamic Crack Growth Criteria In Structural Metals", Workshop on Dynamic Fracture, California Institute of Technology (February 1983) 110–118.
[10] T. Kobayashi and J.W. Dally, in *Crack Arrest Methodology and Applications.* Edited by G.T. Hahn and M.F. Kanninen. ASTM STP 711, American Society for Testing and Materials (1980) 189–210.
[11] J.W. Dally, A. Henzi, and D. Lewis, *Experimental Mechanics* 9, No. 9 (1969) 394–399.
[12] G.R. Irwin, "Series Representation of the Stress Field Around Constant Speed Cracks", University of Maryland Lecture Notes (1980).
[13] S.N. Atluri and T. Nishioka, *Engineering Fracture Mechanics* 18, No. 1 (1983) 1–22.
[14] R.J. Sanford, R. Chona, W.L. Fourney, and G.R. Irwin, "A Photoelastic Study of the Influence of Non-Singular Stresses in Fracture Test Specimens, University of Maryland Report (March 1981).
[15] M. Cottron and A. Lagarde, *SM Archives* 7, Issue 1 (1982) 1–18.
[16] R.J. Sanford and J.W. Dally, *Engineering Fracture Mechanics* 11 (1979) 621–633.
[17] J.F. Kalthoff, "The Shadow Optical Method of Caustics", CISM Lectures (July 1984).
[18] J.W. Phillips and R.J. Sandford, ASTM Special Technical Publication 743, (1982) 387–402.
[19] A.J. Rosakis and K. Ravi-Chandar, "On Crack Tip Stress State: An Experimental Evaluation of Three-dimensional Effects", California Institute of Technology Report, SM 84-2 (March 1984).
[20] W. Yang and L.B. Freund, "Transverse Shear Effects for Through Cracks in an Elastic Plate", Brown University Report (1984).
[21] L.B. Freund, "Some Theoretical Results on the Dependence of the Dynamic Stress Intensity Factor on Crack Tip Speed", Workshop on Dynamic Fracture, California Institute of Technology (February 1983) 129–136.
[22] J. Beinert and J.F. Kalthoff, in *Mechanics of Fracture, Experimental Evaluation of Stress Concentration and Intensity Factors.* Edited by G.C. Sih. Martinus Nijhoff (1981) 281–330.
[23] W. Doll, *Journal of Materials Science* 10 (1975) 935–942.
[24] F. Kerkhof, *Glastechnische Berichte* 33, H. 12 (1960) 456–459.
[25] W.G. Knauss and K. Ravi-Chandar, *International Journal of Fracture* 27 (1985) 127–143.

Résumé

On passe en revue les résultats expérimentaux obtenus par différents chercheurs sur la relation $K_I - \dot{a}$ et l'on discute des différences apparentes dans les résultats qui conduisent à des questions en ce qui regarde l'unicité de cette relation. On présente l'influence des erreurs dues à l'état tridimensionnel des tensions à l'extrémité de la fissure, les effets des contraintes non singulières de la vitesse de la mise en charge transitoire et de la mesure de la vitesse. Ces erreurs ne contribuent pas à éclaircir la question de l'unicité et, en vue de la résoudre, on décrit une expérience possible.

International Journal of Fracture 27 (1985) 169–186.

On the dynamic fracture of structural metals

ARES J. ROSAKIS and ALAN T. ZEHNDER
Graduate Aeronautical Laboratories, California Institute of Technology, Pasadena, CA 91125, USA

(Received September 1, 1984)

Abstract

Some fundamental aspects of dynamic crack growth in structural steels are presented and discussed. The discussion takes the form of a direct comparison of experimental results to elastic-plastic analyses, and attempts to clarify the role of material inertia and plasticity in the dynamic crack growth process. In addition the relation of crack growth criteria to micromechanical void growth models is considered.

Potential problems in the analysis of data obtained by either direct optical measurements or numerical simulations of crack growth are presented. It is demonstrated that large errors in the velocity records caused by stress wave effects are a main source of uncertainty in the interpretation of experimental results.

1. Introduction

For dynamic crack propagation under conditions of small scale yielding, fracture criteria based on the stress intensity factor are expressed in an analogous way to the static situation. This is done by the introduction of the quantity $K_{\mathrm{I}}^{\mathrm{d}}(t)$, which is the time dependent amplitude of the dynamic near tip asymptotic stress field [1], and a critical value K_{Ic} which is usually assumed to be a material property. Since the quantity $K_{\mathrm{Ic}}^{\mathrm{d}}$ represents the resistance of the material to dynamic crack growth, its magnitude can be expected to depend on crack tip velocity as well as on the flow properties of the material. Thus for a propagating crack the fracture criterion is assumed to take the form:

$$K_{\mathrm{I}}^{\mathrm{d}}[a(t), t, \mathrm{Load}] = K_{\mathrm{Ic}}^{\mathrm{d}}(\dot{a}), \tag{1.1}$$

where a, and $\dot{a}$ are the crack tip position and velocity, respectively. The hypothesis expressed by (1.1) states that if the stress-intensity factor, evaluated by a purely elastodynamic analysis, is set equal to the material property function $K_{\mathrm{Ic}}^{\mathrm{d}}$, the dynamic fracture toughness, then the resulting equation of motion will determine $a(t)$ uniquely. The simplicity of the above hypothesis stems from the fact that all nonlinear and plasticity effects are lumped into the material property $K_{\mathrm{Ic}}^{\mathrm{d}}$. This quantity reflects the influence of material inertia on near crack tip nonlinearities and can either be measured by experiments or can be calculated on the basis of nonlinear models of crack growth and more basic, local, crack propagation criteria [2,3]. The question is therefore reduced to whether the fracture toughness $K_{\mathrm{Ic}}^{\mathrm{d}}$ can be safely considered a unique function of crack tip velocity and material properties. The answer, which is still the subject of considerable discussion, seems to depend strongly on the material considered.

2. Limitations of near tip asymptotic solutions

The analytical difficulties associated with the dynamic crack propagation problem in both elastic and elastic-plastic solids have limited the number of available dynamic solutions to idealized situations. In elastic solids, the existing analytical results are valid only until stress waves, reflected from the boundaries, reach the moving crack tip [1]. In ductile materials, there seem to be unresolved differences in the few available asymptotic solutions. In particular, for dynamic plane-strain crack growth in elastic- perfectly plastic materials obeying an associated flow rule, Slepyan [4], Achenbach and Dunayevsky [5] and more recently Gao and Nemat-Nasser [6], obtained three different dynamic asymptotic near-tip solutions. In all cases it was assumed that the yield condition was satisfied at all angles around the crack tip, thus excluding the possibility of a sector of elastic unloading. Achenbach and Dunayevsky used the Tresca yield criterion and concluded that for finite crack tip speeds the strains are bounded whereas a logarithmic singularity is recovered for $\dot{a} = 0$. Also, they showed that for zero crack tip speed and $\nu = 0.5$ the stress field reduces to the Prandtl field. On the other hand Gao and Nemat-Nasser, using the von Mises yield condition, concluded that the strain field varies logarithmically for non-zero crack tip speeds.

Recent finite element work of Lam and Freund [2] on the same problem indicates good agreement between the numerically predicted stress field and the results of Achenbach and Dunayevsky. The numerical work also strongly suggests the possibility of an elastic unloading sector behind the active plastic zone, a result that seems at variance with all available asymptotic analyses. The issue is complicated even further by clear indications that the range of applicability of the asymptotic results diminishes very rapidly and eventually disappears as the crack tip velocity vanishes. This has been clearly demonstrated for the case of mode-III elastic-plastic crack growth by Freund and Douglas [3]. By using a full field analytic solution they were able to demonstrate that the equivalent dynamic asymptotic result due to Slepyan [4] has a validity range of less than a hundredth of the plastic zone (process-zone size) for crack tip velocities less than 0.3 C_s. This limitation which is also expected to apply, at least qualitatively, to the mode-I asymptotic dynamic solutions presents a major drawback in the applicability of the analytical results. Such uncertainties and limitations make numerical methods and direct experimental observations necessary for the complete understanding of the dynamic fracture event.

3. The effect of inertia on dynamic fracture toughness

3.1. Analytical and numerical predictions

For crack propagation in elastic-plastic solids under conditions of contained yielding, (most structural steels) some theoretical speculations can be made about the nature of the dependence of the dynamic fracture toughness on crack tip speed. It can be argued from a purely intuitive point of view that in the fracture of rate independent elastic-plastic materials the plastic zone surrounding the tip will give rise to greater inertia forces as the velocity of crack propagation increases and also that the effect will be much larger than in purely elastic materials. Approaching the problem from both an analytical and numerical point of view, Freund and Douglas [3] and Douglas [7] were able to study the steady state motion of mode-III cracks, propagating dynamically in an elastic, perfectly-plastic solid. The material was characterized by the von Mises yield criterion and a J_2 flow theory of plasticity. The results were concerned both with the full deformation field, which was determined by means of the finite element method, and with the deformation on the crack line within the active plastic zone, which was determined analytically. Although the full field numerical analysis was conducted under the assumption of small scale yielding, the analytical result did not depend on this restriction. The main observation on the strain distribution was that the level of plastic strain is significantly reduced from its corresponding slow crack growth levels due to material inertia. If this result is combined with the requirement of a fixed level of plastic strain at a critical distance in front of the crack tip (a criterion proposed by McClintock and Irwin [8]) it can be expected that, for crack growth to occur, the far field stress intensity factor would necessarily increase with increasing crack speed. To quantify this idea Freund and Douglas [3] related the remotely applied stress intensity factor to the plastic zone strain distribution through the full-field numerical solution. The result displayed here in Fig. 1, demonstrates the variation of K^{d}_{IIIc} with crack tip velocity. The parameters appearing in Fig. 1 are the elastic shear wave speed c_s, the critical plastic strain γ_f and the level of applied stress intensity K_{IIIc} required to satisfy the same fracture criterion for crack initiation in the same material.

With a similar view toward developing a theoretical relation between K_{Ic} and crack tip speed, Lam and Freund [2] have analyzed the elastic-plastic, mode-I problem. In their work the stress intensity factor was related to the near tip crack opening displacement through the full field numerical solution. A critical crack tip opening angle growth criterion was imposed which was similar to that used by Dean and Hutchinson [9] and by Lam and McMeeking [10]. As in the mode III problem, the approach was based on the requirement that the same growth criterion can be applied to both initiation and crack propagation in the same material. The results are presented here in Fig. 2. The figure

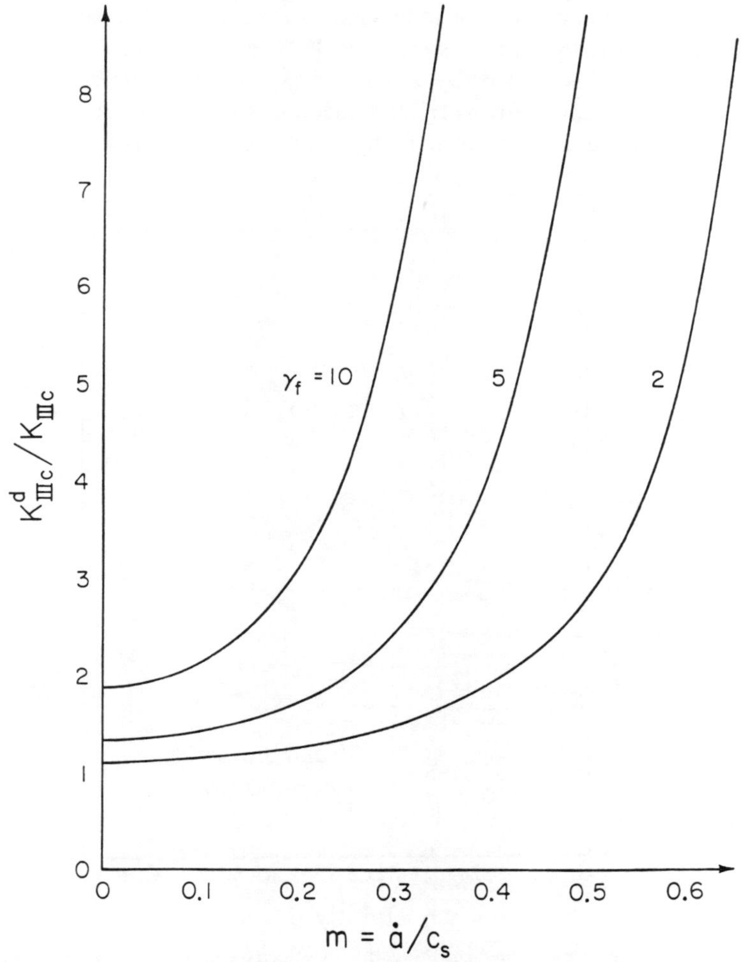

Figure 1. Mode III dynamic fracture toughness as a function of crack tip velocity in an elastic-ideally plastic rate insensitive material, according to the critical plastic strain fracture criterion. (From Freund and Douglas [3].)

shows the variation of K_{Ic}^{d}/K_{Ic} with normalized crack speed for different values of the ratio δ_c/r_m where δ_c is the critical value of the crack opening displacement at a characteristic distance r_m measured from the crack tip along the crack faces.

The importance of this result is amplified by the possibility of direct comparison with experiments performed in metals that fracture with a locally ductile mechanism while preserving the requirements of small scale yielding. Indeed, the numerical prediction corresponding to the value of $\delta_c/r_m = 32$ seems to agree well with the experimental data for dynamic crack growth in an AISI 4340 steel which was reported by Rosakis [11] and Rosakis, Duffy, and Freund [12,13] which data are also included in Fig. 2. The measurements were performed by means of the optical method of reflected caustics with a double cantilever beam specimen in a wedge loaded configuration. The data points represent the cumulative result for three identical specimens. As explained in detail in [2], knowledge of the material properties of this particular 4340 carbon steel (heated to 843°C, oil quenched and tempered at 316°C for an hour, $\sigma_0 = 1300$ MPa, $K_{Ic} = 30$ MPA$\backslash\sqrt{m}$) dictate the choice of $\delta_c/r_m = 32$ and allow the direct comparison of the experimentally obtained curve to this particular member of the K_{Ic}/K_{Ic} vs. $\dot{a}/c_s$ family.

It should perhaps be noted at this point that the crack growth in the dynamic experiments was not steady state whereas the numerically obtained curves were derived using the steady state assumption. Thus direct comparison between the two results is possible only if the dynamic fracture toughness is a weak function of crack tip acceleration. Experiments reported recently by Brickstad [14] and described in detail in Section 3.2 (Fig. 3) indicate that this is indeed the case for some structural steels.

Local crack initiation and growth criteria such as the ones described above, represent attempts to model some of the basic features of the micromechanics of the fracture

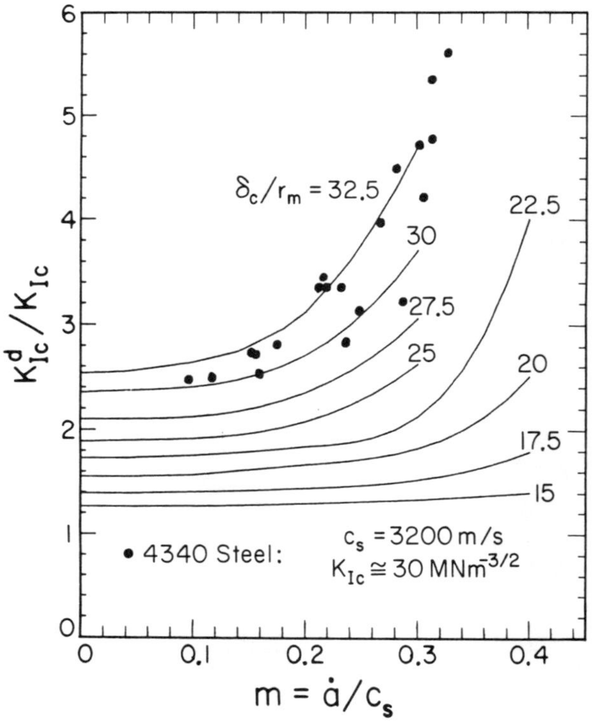

Figure 2. Mode-I dynamic fracture toughness as a function of velocity according to the critical crack tip opening angle fracture criterion. Also included are experimental results of Rosakis, Freund and Duffy [13] for 4340 steel. (From Lam and Freund [2].)

process. The fracture of ductile metals has been observed to occur by the initiation, growth, and coalescence of voids (Beachem [15] and Rosenfield [16]). Such voids are generated by plastic flow in a zone of intense straining surrounding the crack tip, called the process zone. Requiring that a certain displacement or strain reaches a critical value at a characteristic distance away from the tip is equivalent to identifying this characteristic size with the process zone. Inside this zone, the continuum theory is no longer strictly applicable. Thus, the development of comprehensive fracture criteria requires a detailed study of void growth under the macroscopically imposed crack tip field. The mechanism of quasistatic plastic hole growth was initially studied by McClintock [17] and by Rice and Tracey [18]. Both investigations emphasized the important role of triaxial stress states in enhancing void growth in the near-tip region. As pointed out by Rice [19], the high near-tip triaxiality can nucleate voids from preexisting inclusions or impurity particles and can lead to a macroscopically brittle crack behavior in metals which on the microscale are significantly ductile. An example of such behavior can be observed in the particular heat treatment of the AISI 4340 steel tested in [13].

For dynamic crack growth in metals which fracture by the locally ductile mechanism of hole growth under small scale yielding conditions, as in the cases described above, quantitative estimates of the influence of material ductility and inertia on the relationship between K_{IC}^{d} and $\dot{a}$ can be made. Glennie [20] modified the void growth formulation of Rice and Tracey to include the effect of inertia, and obtained an approximate solution by means of a variational method. More recently Rosakis, Duffy, and Freund [13], following a similar approach, used an elementary argument to estimate the relative influence of inertia in the near tip ductile hole growth process. They showed that the total energy required for dynamic hole growth in front of a propagating crack tip, (kinetic energy plus dissipation through plastic work), normalized by the plastic work, is an increasing function of crack tip speed. The sensitivity of this ratio to crack tip speed was found to depend on the flow stress of the material. It was demonstrated that for more ductile solids,

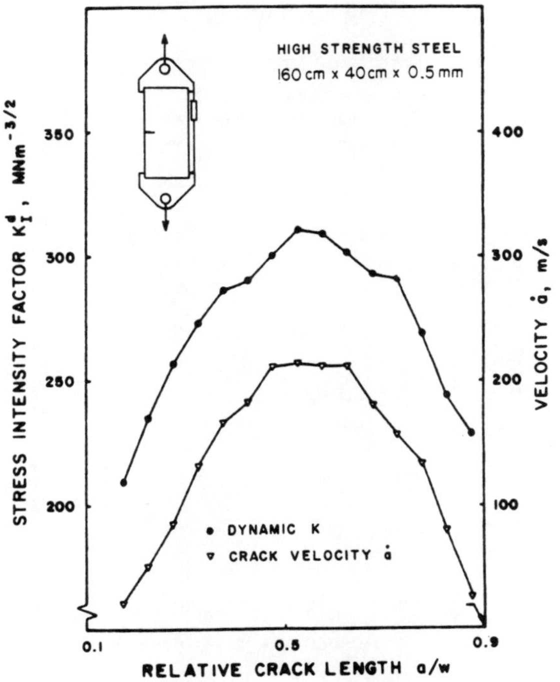

Figure 3. Stress intensity factor and crack velocity as functions of crack length. (From Brickstad [14].)

the inertia effects become important at lower speeds than in brittle solids, a behavior consistent with the results displayed in Fig. 2.

3.2. Experimental predictions

The available experimental results in the dynamic fracture of metals are limited to crack propagation under small scale yielding conditions. The principal purpose of such investigations is to observe the instantaneous stress intensity factor, and to determine whether or not the observed variation constitutes a material characteristic. The experiments can be divided into boundary value and direct measurements.

Boundary value measurements

The boundary value techniques involve measurement of crack tip position vs. time and either measurement of the time-dependent load applied to the specimen (Brikstad [14], Kanazawa et al. [22]) or an assumption regarding the nature of the boundary conditions (Bilek [23]), i.e. fixed displacement or constant rate of displacement at the load points. The results of such measurements are used in conjunction with a dynamic numerical model for the prediction of the stress intensity factor history.

Along these lines, Brickstad tested SEN steel specimens loaded in tension. The time dependent boundary conditions were measured with strain gages and the crack tip speed was recorded by using the R.F. current technique developed by Carlsson [24], to obtain a continuous recording of crack length. The energy release rate, and hence K_I^d, was then obtained by calculating the work done by the closing forces at the crack tip by means of a dynamic FEM program. By using a stretching screw on the side of the specimen away from the machined precrack, an initially increasing and subsequently decreasing K field was obtained. This produced both acceleration and deceleration phases of crack growth and allowed the study of the sensitivity of the dependence of the fracture toughness on $\ddot{a}$. Results from a single specimen are presented in Fig. 3 where both K_I^d and $\dot{a}$ are given as a

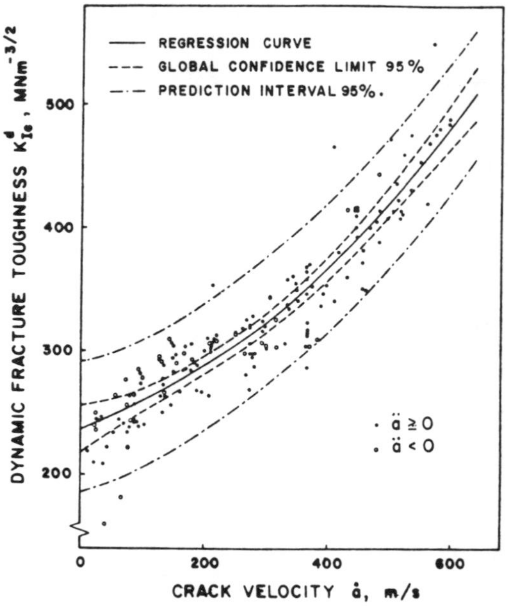

Figure 4. Dynamic fracture toughness as a function of crack velocity for a high strength steel. (From Brickstad [14].)

function of crack length. As is obvious from this figure, the crack tip velocity and K_I^d vary in phase both in the acceleration and deceleration regimes of the crack growth history, clearly indicating a relation between the two quantities. To demonstrate this more clearly, the collective data from many specimens is presented in Fig. 4 where a clear relation between K_{Ic}^d and $\dot{a}$ is indicated while no dependence on acceleration is observed, contrary to previously reported work [25].

Behavior very similar to the above was also reported by Bilek [23] who performed tests with DCB specimens of 4340 steel. In this work the crack tip speed was measured by the same R.F. current technique used in [14] and also by crack propagation (ladder) gages with a spacing of 0.2 cm. A dynamic, Timoshenko beam model of the DCB specimen was then used to infer the dynamic stress intensity factor history. Despite the large scatter due to the many different tests being presented, the variation of K_I with crack tip speed, illustrated here in Fig. 5 follows the same trend as in [13] and [14] and compares well with the superimposed results of Hahn et al. [27] and Angelino [28]. Angelino devised a unique procedure for the determination of K_I^d and the crack tip speed. Using small, three point bend specimens, a crack was propagated unstably for small distances. The material used was a carbon steel very similar to 4340. By measuring the crack length and time of propagation (about 7 μs) the average velocity was obtained. Measurement of the strain energy in the specimen before and after crack propagation gave the fracture energy, which when divided by the fracture surface area yielded the energy release rate.

Additional confidence in the existence of a unique K_{Ic}^d vs. $\dot{a}$ relation in steel is obtained by Kanazawa et al. [22]. They performed experiments using both DCB and large SEN specimens subjected to uniform tension. The crack tip position was recorded by the use of gages spaced 3 cm apart. This information produced an *average* velocity record that was used in conjunction to a dynamic finite different code. By computing the energy variation, the dynamic energy release rate was obtained by a global energy balance in the specimen. Of particular interest are results corresponding to SEN specimens tested with a linear temperature gradient varying from −100°C on one side to +50°C on the other, the temperature increasing with crack length. As the crack propagated into the higher temperature region it decelerated and eventually stopped. The family of K_{Ic} vs. $\dot{a}$

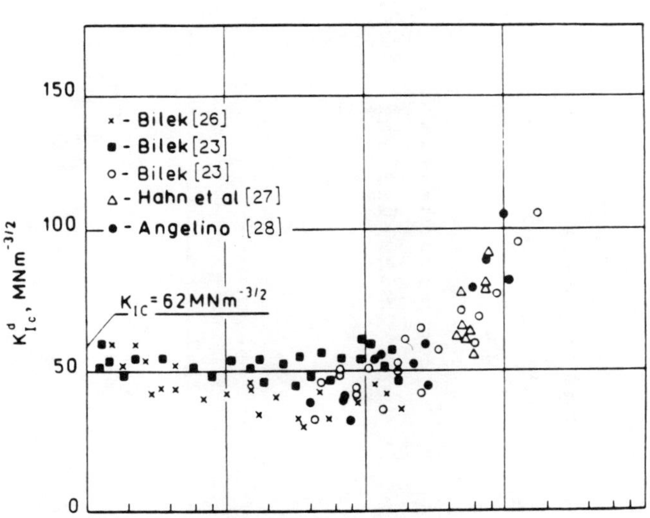

Figure 5. Dynamic fracture toughness as a function of crack velocity for 4340 steel. (From Bilek [23].)

relations presented in Fig. 6, corresponding to different temperatures, can explain the observed deceleration behavior of the cracks. The fracture toughness $K_{\mathrm{Ic}}^{\mathrm{d}}(\dot{a})$ increases with temperature (and hence crack length in these tests), however the stress intensity factor $K_{\mathrm{Ic}}^{\mathrm{d}}$ available was nearly constant, thus to satisfy (1.1) the crack must slow down until $K_{\mathrm{Ic}}^{\mathrm{d}}(\dot{a}) = K_{\mathrm{I}}^{\mathrm{d}}$. Cumulative results for $-40°$C are given in Fig. 7. It has been theorized [29] that the vertical bands of constant speed shown in the high velocity region of this curve are due to non-uniqueness in the $K_{\mathrm{Ic}}^{\mathrm{d}}$ vs. $\dot{a}$ relation, and is similar to the behavior observed in Homalite-100 [30]. The authors would like to point out that in the above tests the wide gage spacing (3 cm) only gives an averaged record of crack tip speed. Thus an alternative explanation is that fine variations in the velocity are not detectable, leading to different values of $K_{\mathrm{I}}^{\mathrm{d}}$ being assigned to the same average velocity (vertical lines).

Direct optical measurements

For the investigation of crack growth in transparent materials the techniques of photoelasticity and transmitted caustics have been used extensively in conjunction with high speed photography. Through application of these techniques, important advances in the study of rapid fracture have been achieved particularly by Dally [31], by A.S. Kobayashi [32,33] by Kalthoff and coworkers [34] and by Ravi-Chandar and Knauss [21]. These methods have been extended to metals through the use of photoelastic coatings and reflected caustics. Other techniques such as dynamic interferometry, moiré and the newly reported stress intensity factor tracer technique, are currently being explored [35,36]. As expected, the number of direct measurements performed on metallic fracture specimens is small compared to ones in clear polymers or glasses. This is due to the added technical difficulties introduced by reflected high speed photography and to the high level of effort associated with specimen preparation.

Birefringent coatings were first used in the study of the dynamic fracture of steel by T. Kobayashi and Dally [37]. As discussed in their work, the use of a single coating covering the surface of the metal specimen raises questions concerning the relative position of the crack tip, in the plastic coating and in the metal specimen. However, T. Kobayashi and Dally overcame this problem by cementing a pair of coatings to the surface of the grooved

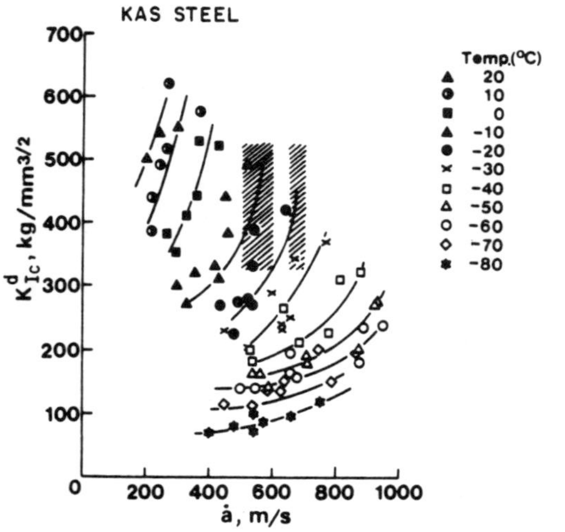

Figure 6. Dynamic fracture toughness as a function of crack velocity for different temperatures. 100 kg mm$^{-3/2}$ = 31 MNm$^{-3/2}$. (From Kanazawa et al. [22].)

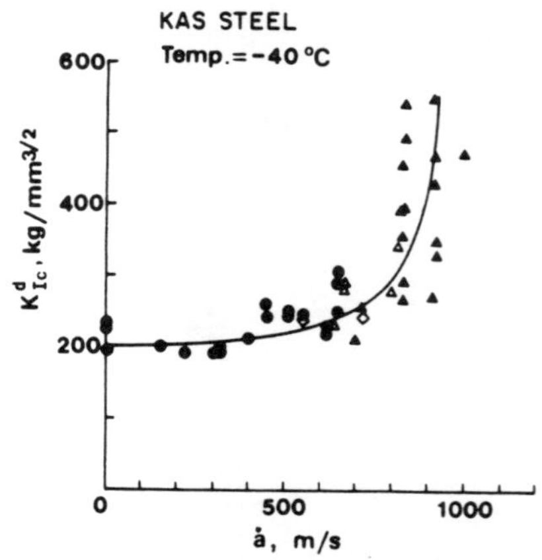

Figure 7. Dynamic fracture toughness as a function of crack velocity at $-40°$C. (From Kanazawa et al. [22].)

4340 steel specimen, one on either side of the groove. As the crack advanced they recorded the isochromatic patterns using a Cordin high-speed framing camera, from which they calculated the stress intensity as a function of time. They were also able to determine crack length and hence the crack velocity at each position. The results indicate large oscillations in the value of K_I^d corresponding to abrupt changes in the crack tip velocity. As discussed in detail in [12], the oscillations in K_I^d and in crack speed are in phase and are caused by the interaction of reflected stress waves with the propagating crack tip. Results from three different specimens are presented in Fig. 8 where the variation of K_{Ic}^d with $\dot{a}$ is illustrated. The results follow the same qualitative trend as those reported by Bilek for a similar heat treatment of the 4340 steel. The specimen labeled 375 gave anomalous results, which were attributed by the authors to a different heat treatment.

The method of caustics (shadow spot method) for transparent materials was introduced by Manogg [38]. Through such contributions as those of Theocaris and Gdoutos [39], and Beinert and Kalthoff [34] this method has become established as a standard procedure for the measurement of the static and dynamic fracture toughness of nominally elastic, transparent materials. In addition, within the past few years, Rosakis [40] and Rosakis and Freund [41] provided analyses to account for the effects of material inertia and crack tip plasticity in the interpretation of caustic data obtained with light reflected from specimens of opaque structural materials. Extensive reviews of the method are given by Beinert and Kalthoff [34] and more recently by Rosakis and Zehnder [42] who presented the exact equations of caustics and discussed the consequences of mathematical approximations made by previous analyses.

The method was first applied to the study of the arrest process in high-strength steels, by Beinert and Kalthoff [34], but a relation between K_{Ic}^d and $\dot{a}$ was not reported. More recently Rosakis [11], and Rosakis, Duffy, and Freund [12,13] used reflected caustics to study the dynamic behavior of cracks propagating rapidly in double cantilever beam specimens of 4340 steel. The caustic patterns formed during the course of the experiments were recorded by means of a high speed camera of the Cranz-Schardin type. The instantaneous value of the dynamic stress intensity factor and the crack tip position were thus recorded. The results of the experiments indicate that the dynamic fracture toughness is an increasing function of crack tip speed (see Fig. 2). These values exceed those of

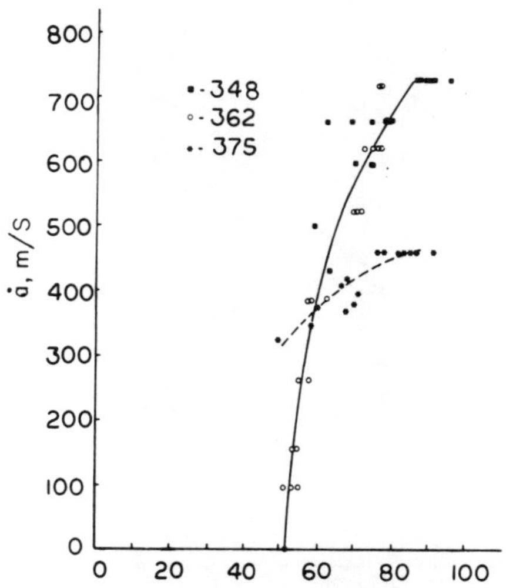

Figure 8. Dynamic fracture toughness as a function of crack velocity for a 4340 steel. (From Kobayashi and Dally [37].)

T. Kobayashi and Dally and those of Bilek et al. by a consistent 10%. This difference can easily be attributed to different mechanical properties due to differences in heat treatment. Of greater significance than the absolute values of K_{Ic}^{d} is the general shape of the curve and the qualitative agreement with the numerical results of Lam and Freund as demonstrated in Fig. 2.

A second feature of these results that is consistent with the observations of T. Kobayashi and Dally are the oscillations in the time records of K_{Ic}^{d} and $\dot{a}$. These oscillations were also correlated with the arrival time of stress waves that were initially emitted from the accelerating crack tip and subsequently reflected from the DCB specimen boundaries. Recent results obtained by Rosakis and Zehnder [43], presented here in Figs. 9 and 13a, also display such oscillations. Shown in Fig. 13a are the arrival times of waves reflected from the lateral sides, loading pins and the end of the cantilever arms of the DCB specimen. A detailed discussion on the influence of the reflected waves on dynamic crack advance can be found in [12 and 13]. Visual evidence of the existence of unloading waves generated at the propagating crack tip are shown in Fig. 10a. The photograph was taken by the present authors with the experimental setup sketched in Fig. 11. The photograph shows the caustic image of a dynamic crack ($\dot{a} = 1000$ m/s) and also the shadow images of unloading waves generated at the tip. The waves were found to propagate with the Rayleigh wave speed. It is perhaps worth noting here that only waves that produce out of plane displacements on the polished specimen surface can be detected by the caustics setup. Under idealized plane stress conditions this limits the possibilities to either Rayleigh waves propagating on the plate surface, or dilatational waves which produce a lateral contraction of the planar specimen due to the Poisson ratio effect.

Plane stress has been a key assumption in the interpretation of results from caustics

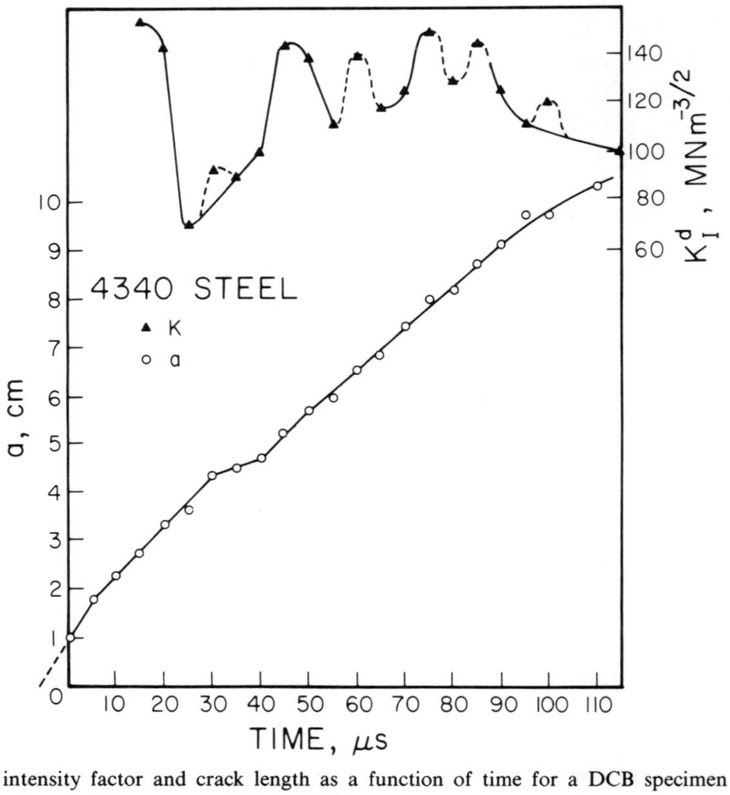

Figure 9. Stress intensity factor and crack length as a function of time for a DCB specimen of 4340 steel. (Rosakis and Zehnder, work in progress, Caltech 1984.)

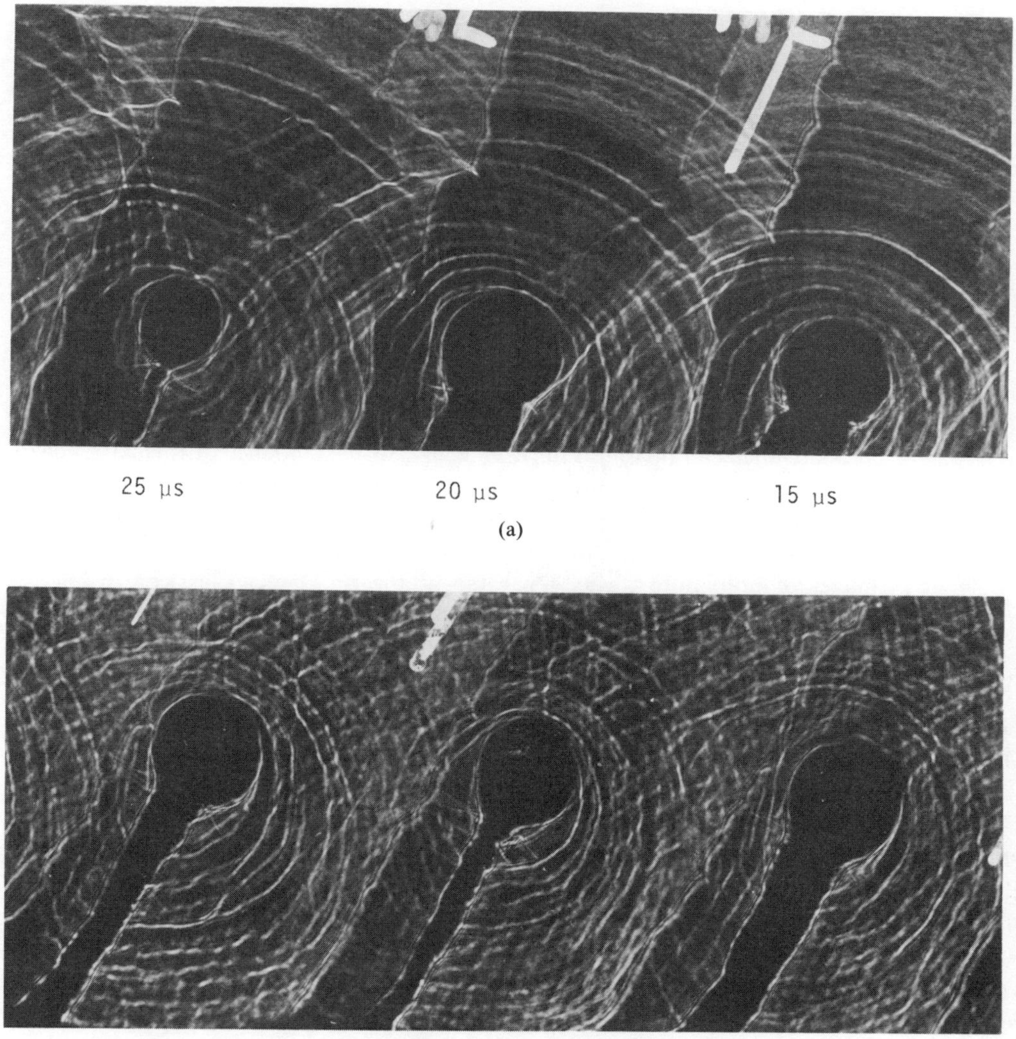

25 μs 20 μs 15 μs

(a)

60 μs 55 μs 50 μs

(b)

Figure 10. Reflected caustic patterns corresponding to dynamic crack growth test recorded in Fig. 9 for 4340 steel. (a) Strong dynamic effects are demonstrated by visible unloading waves and oscillations in caustic diameter. (b) Visible oscillations in the width of wake indicate vibrations of the cantilever arms of DCB specimens. (Rosakis and Zehnder, work in progress, Caltech 1984.)

data. In particular it has been assumed that at distances sufficiently far from the crack tip, the stress and displacement fields are adequately described by the theory of plane stress elasticity. Under the action of the applied loads, the specimen undergoes nonuniform out-of-plane deformation (thinning) whose intensity is completely characterized by the elastic plane-stress intensity factor.

With a view towards establishing the limitation of the plane stress interpretation of caustics data, Rosakis and Ravi-Chandar [44] tested compact tension specimens with sharp, planar straight-fronted cracks. The specimens were loaded to a fixed level and then shadow spot data were taken at different distances from the crack tip. This was achieved by adjusting the optical arrangement used in the experiment. Some of their results are

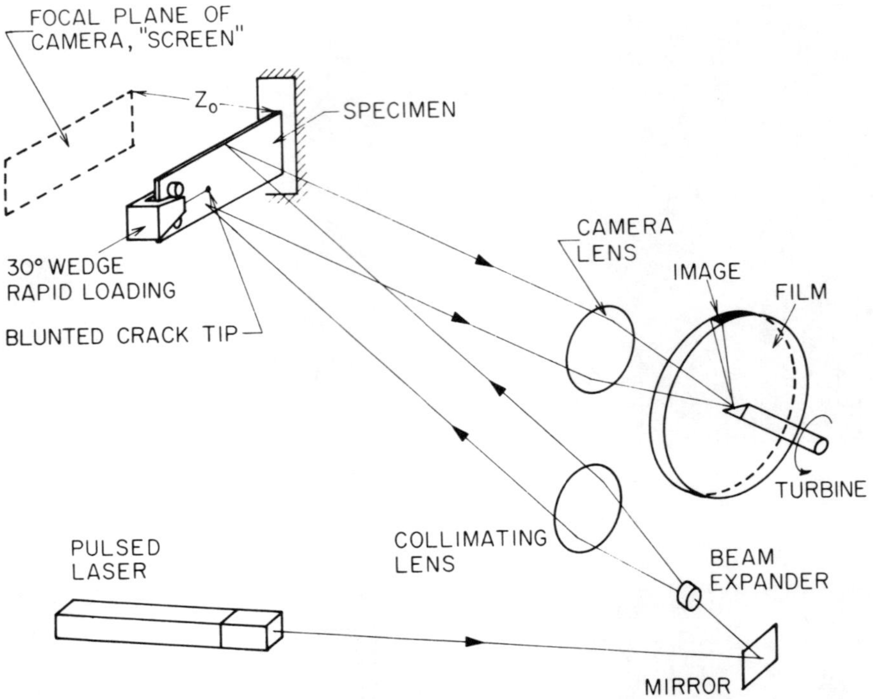

Figure 11. Experimental set-up for reflected, high speed photography of caustics. (Rosakis and Zehnder, work in progress, Caltech 1984.)

shown in Fig. 12. The figure contains data for martensitic 4340 steel specimens of several thicknesses. The ordinate is the ratio of the optically determined stress intensity factor (at different measurement distances from the crack tip) to the corresponding plane stress value obtained by boundary value measurements. The abscissa is the distance r from the crack tip (where the measurement was performed) normalized by the specimen thickness h. The results indicate that plane-stress conditions prevail at distances from the crack tip greater than half the specimen thickness. This conclusion is consistent with the crack tip

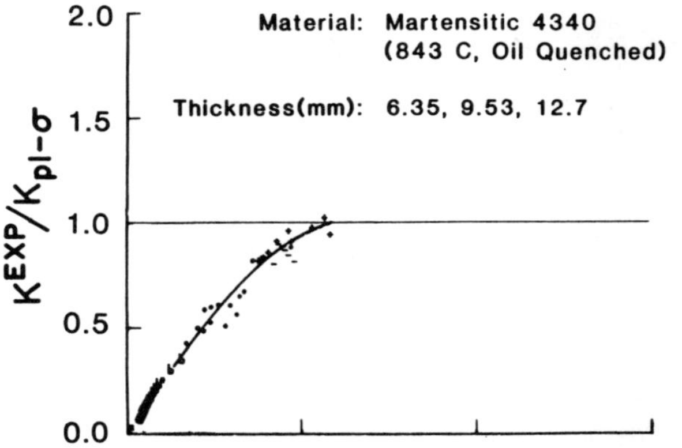

Figure 12. Static stress intensity factor as determined by caustics, divided by plane stress value for different distances, r, from the crack tip. (From Rosakis and Ravi-Chandar [44].)

boundary layer solution of Yang and Freund [45] and with the numerical results of Levy, Marcal, and Rice [46]. No such analysis exists for the dynamic problem. Nevertheless, using the static case as a guideline, the present authors feel that accurate optical measurements based on two-dimensional theory (photoelasticity, interferometry, caustics) can be performed with confidence only at distances greater than one half the specimen thickness.

Plasticity at the crack tip also limits the minimum distance from the tip where optical measurements can be performed. Analysis by Rosakis and Freund [41] has shown that the error introduced through the neglect of plasticity effects in the interpretation of caustics data will be small as long as the measurement is performed in a radius greater than twice the plastic zone size. Also, for any extent of the plastic zone, inertia effects appear to be significant for crack speeds exceeding approximately 0.2 c_1, where c_1 is the longitudinal wave speed. Analysis of caustics formed by reflection of light from within the plastic zone exists only for stationary cracks [47].

4. Techniques of velocity measurement and inaccuracies due to errors in the velocity records

Accurate measurement of instantaneous crack tip speed is both very difficult and very important. The large scatter seen in $K_{\mathrm{IC}}^{\mathrm{d}}$ vs. $\dot{a}$ data such as Figs. 5 and 7 is most likely due to inaccurate crack tip position records resulting in poor velocity calculations. Several techniques, which can be divided into continuous and discrete are mentioned here and discussed. For a more thorough review see Weimer and Rogers [48].

One of the more widely used continuous measurements is the radio frequency skin effect technique developed by Carlsson [24]. High frequency currents tend to travel on the surface of a conductor, a behavior commonly referred to as the skin effect. By positioning terminals on either face of the crack, the path traveled by the current and thus the effective impedance increases linearly with crack length. Recording the impedance then gives a continuous record of the crack length history. An alternative method based on the overall change in the resistance of a metal specimen, due to crack growth, was developed by Congleton and Denton [49]. It was found that the change in potential of a direct current passed from one side of a steel SEN specimen to the other can be used to obtain a continuous record of crack tip position.

Discrete measurements involve either high-speed photography, as in photoelasticity and caustics, or a grid of conductive material of some type placed on the specimen along the prospective crack path [23,49,50]. As the crack runs, the conducting strips are broken and the resistance of the whole grid is increased. This stepwise change in resistance can be observed by using a bridge circuit and a recording oscilloscope. The crack length $a(t)$ is then known and can be differentiated to yield the crack tip velocity. Similarly, measurements of crack length at fixed time intervals are taken with high speed photography.

If the interval of measurement is not small enough, large errors may result due to variations in velocity on time scales smaller than the time interval. This can be expected to cause problems in small specimens where reflections of stress waves from the boundaries may cause abrupt changes in crack tip position. The $a(t)$ record is then subject to interpretation. For example, consider the $a(t)$ records in Fig. 13 obtained with the aid of caustics in experiments performed by the authors on 4340 steel. For this record data were obtained at intervals of 7 μs. In Fig. 13b a smooth curve is drawn through the points producing one $\dot{a}(t)$ record. In Fig. 13a kinks are allowed in the $a(t)$ record at points where the first reflected stress waves arrive at the crack tip. This is done by fitting smooth curves to the right and left of the wave arrival time and then matching the curves at the wave arrival, allowing a discontinuity in slope. As can be seen the second interpretation produces a much different $\dot{a}(t)$ record. As a result of the above, the authors feel that in

(a)

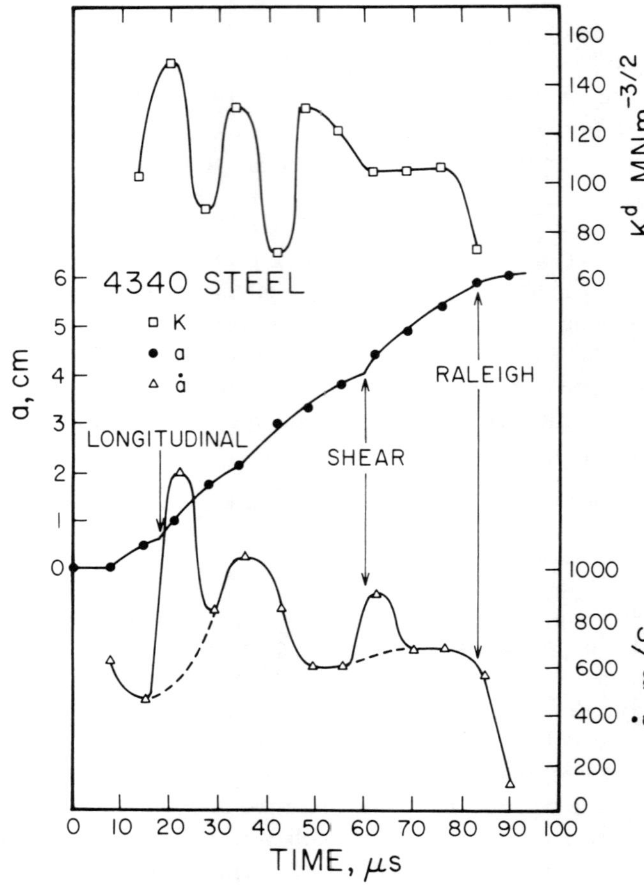

(b)

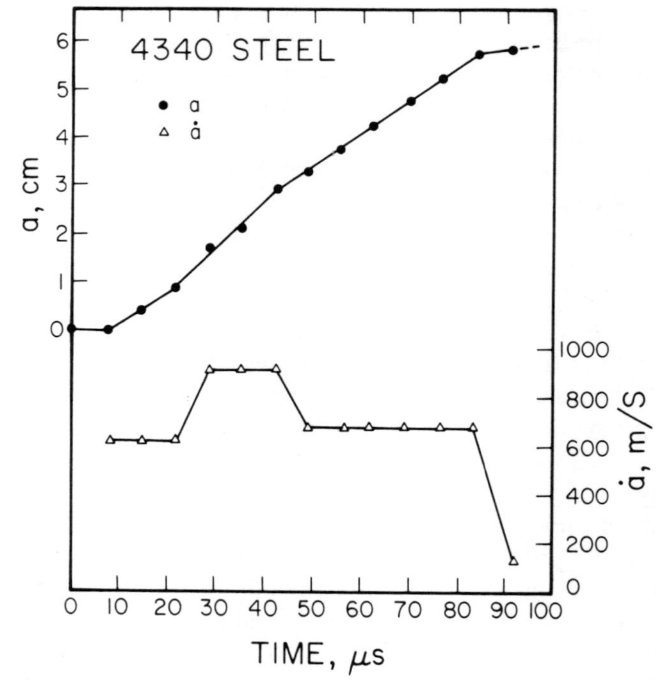

Figure 13. (a) Stress intensity factor, crack length and crack velocity as functions of time. Arrows indicate reflected stress wave arrivals. (b) Alternative interpretation of $a(t)$ record. (Rosakis and Zehnder, work in progress, Caltech 1984.)

order to achieve accuracy in $\dot{a}(t)$, the crack tip position must be measured at intervals less than 5μs, preferably at least every 2 μs. This is particularly true for metals where the typical terminal velocities are three times higher than in polymers.

Some consequences of velocity record uncertainty were investigated by Hodulak, A.S. Kobayashi and Emery [51] who demonstrated the sensitivity of computed K_I^d values to crack velocities used as imput to FEM programs. For wedge loaded RDCB specimens of Araldite B, they used the measured $a(t)$ record to compute $K_I^d(t)$. They then perturbed the experimental $a(t)$ and recalculated $K_I^d(t)$. Their results show that small errors in velocity can severely perturb the predicted value of K_I^d.

Further indications of this can be seen in the work of Nishioka and Atluri [52]. They performed a FEM simulation of Kalthoff's experiments in 9 mm thick wedge loaded RDCB specimens of high strength steel [53]. Kalthoff's $K_I^d(a)$ and $\dot{a}(a)$ records are given in Fig. 14 together with the numerically predicted values. Note that the calculated K_I^d does not show the oscillations that Kalthoff et al. report and attribute to wave interactions. A possible explanation of the discrepancy may lie in the fact that an averaged velocity record was used. In particular, for the first 200 mm, a straight line was drawn through the $a(t)$ data, resulting in constant velocity, whereas for longer crack lengths the velocity is not averaged. As evident from the figure, it is in this region ($a > 200$ mm) that the calculated K_I^d agrees with the measured values. The work of Hodulak et al., indicates the possibility that the disagreement between the measured and calculated K_I^d is due to not inputting accurate velocity measurements to the FEM analysis.

Further consequences of uncertainties in the velocity record, in the presence of stress wave reflections, are also demonstrated by Dahlberg, Nilsson and Brickstad's [54] results reproduced here in Figs. 15a, b. They used SEN specimens loaded in uniform tension. The results in Fig. 15a are for specimens 60 cm high, those in Fig. 15b are for specimens 160

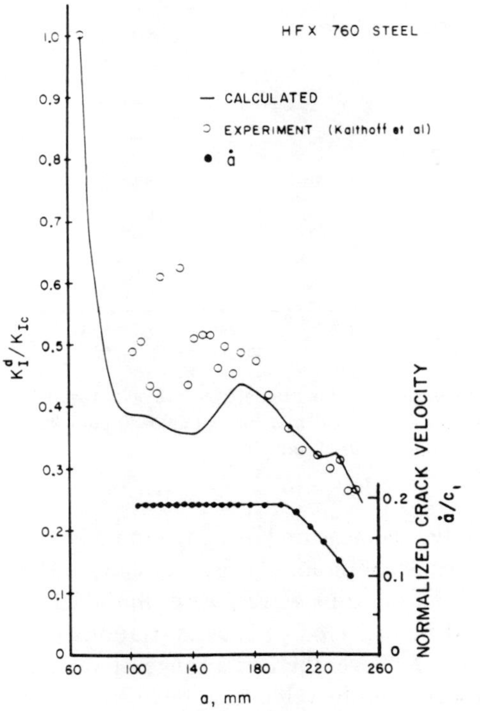

Figure 14. Experimentally determined crack velocity and stress intensity factor (from Kalthoff et al. [53]) and computed stress intensity factor (from Nishioka and Atluri [52]).

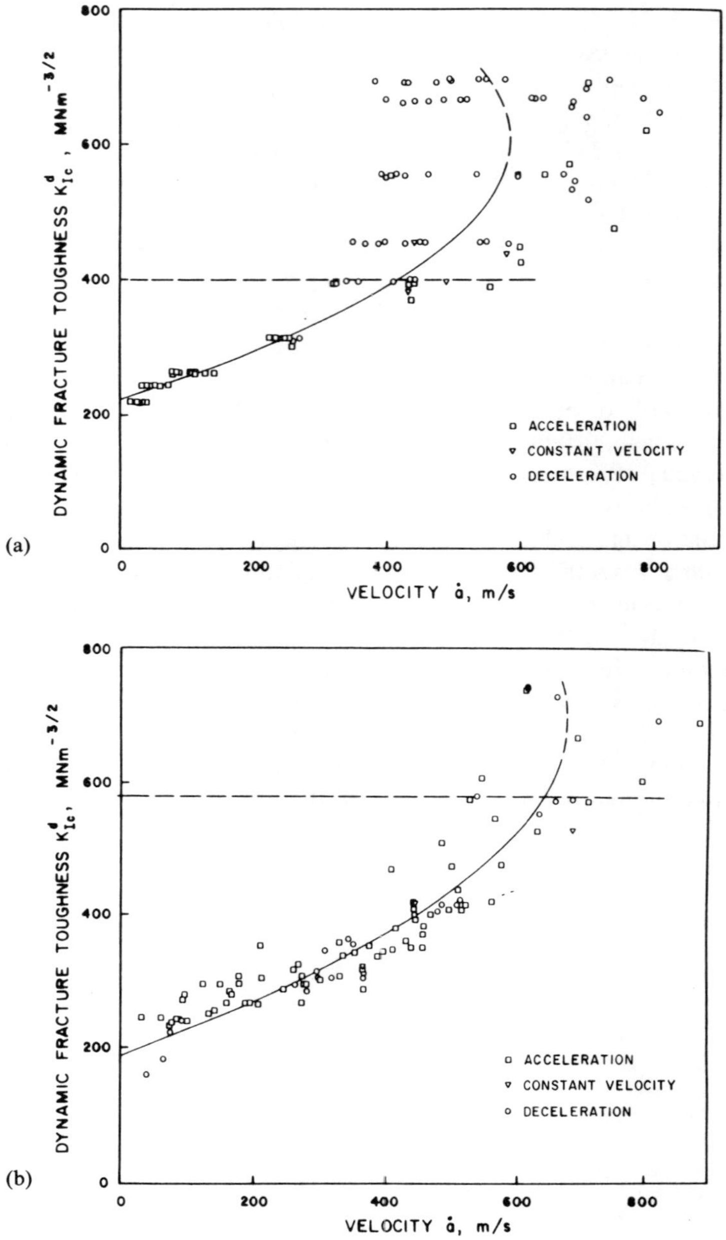

Figure 15. Dynamic fracture toughness as a function of crack velocity for SEN specimens of high strength steel. Specimen dimensions, (a) 60 cm × 40 cm × 0.5 mm (b) 160 cm × 40 cm × 0.5 mm. Increased scatter in smaller specimen. (From Dahlberg, Nilsson, and Brickstad [54].)

cm high. The crack velocity was measured using the impedance technique discussed earlier in this section which gives a continuous measure of $a(t)$. However, oscillations in $a(t)$ whether from electrical noise or stress waves, were smoothed out. In the smaller specimen where stress waves would be expected to cause variations in $\dot{a}$, there is much scatter in data. Note that for a given K_I^d level there is a range of velocities. This seems likely to be caused by experimental scatter in the velocities due to stress wave reflections. On the other hand, in the larger specimen, where reflected stress waves are *not* present, there is less scatter in the K_I vs $\dot{a}$ data.

Evidence that stress waves do cause abrupt variations in K_I^d in DCB specimens is seen in Figs. 10a, b, photographs corresponding to the test recorded in Fig. 9. In Fig. 10a the width of the wake behind the caustic oscillates indicating oscillations in the beam arms of the DCB specimen. In Fig. 10b the caustic diameter is suddenly reduced from one frame to the next (5 μs), due to the arrival of an unloading stress wave. This direct evidence of large oscillations in the DCB specimen cantilever arms clearly demonstrates that fixed grip conditions are not a good assumption for the numerical modelling of DCB specimens.

5. Concluding remarks

The qualitative agreement of results from three entirely different points of view (continuum models, experiment and microscopic void growth models) gives confidence in the use of the local, ductile, crack growth criteria discussed here. It also provides strong evidence for the existence of a unique dependence of K_{Ic}^d on crack tip speed in metals that fracture with a locally ductile mechanism of hole growth. For brittle solids the separation process seems to be stress controlled (generation of microcracks) and the above discussion is no longer applicable. This is demonstrated by the recent experimental results of Ravi-Chandar and Knauss [21] which indicate a non-unique K_{Ic}^d vs. $\dot{a}$ variation in Homalite-100. However, no quantitative explanation, in terms of a microscopic fracture-process model, is yet available.

Acknowledgements

The authors would like to express their gratitude to Dr P.S. Lam and Professor L.B. Freund of Brown University for providing them with a copy of their work as well as with results displayed in Fig. 2, prior to publication of their work. The authors would also like to acknowledge the laboratory assistance of Mr. Sridhar Krishnaswamy. The research support of the National Science Foundation, Grant No. MEA-83-07785, as well as the support of an IBM (faculty development award) is also gratefully acknowledged.

References

[1] L.B. Freund, in *Mechanics of Fracture,* ASME-AMD 19 (1976) 105–134.
[2] P.S. Lam and L.B. Freund, *Journal of the Mechanics and Physics of Solids* (1984) to appear.
[3] L.B. Freund and A.S. Douglas, *Journal of the Mechanics and Physics of Solids* 30 (1982) 59–74.
[4] L.I. Slepyan, *Mekhanika Tverdogo Tela* (English translation) 11 (1976) 144–153.
[5] J.D. Achenbach and V. Dunayevsky, *Journal of the Mechanics and Physics of Solids* 29 (1981) 283–303.
[6] Y.C. Gao and S. Nemat-Nasser, *Mechanics of Materials* 2 (1983) 47–60.
[7] A.S. Douglas, Ph.D. Thesis, Brown University (1981).
[8] F.A. McClintock and G.R. Irwin, in *Fracture Toughness Testing and its Applications,* ASTM-STP 381 (1964) 95.
[9] R. Dean and J.W. Hutchinson, in *Fracture Mechanics: Twelfth Conference,* ASTM-STP 700 (1980) 383–405.
[10] P.S. Lam and R.M. McMeeking, *Journal of the Mechanics and Physics of Solids* (1984) 395–414.
[11] A.J. Rosakis, Ph.D. Thesis, Brown University (1982).
[12] A.J. Rosakis, J. Duffy and L.B. Freund, in *Workshop on Dynamic Fracture Proceedings.* Edited by W.G. Knauss, K. Ravi-Chandar and A.J. Rosakis, California Institute of Technology (1983) 100–118.
[13] A.J. Rosakis, J. Duffy and L.B. Freund, *Journal of the Mechanics and Physics of Solids* 32 (1984) 443–460.
[14] B. Brickstad, *International Journal of Fracture* 21 (1983) 177–194.
[15] C.D. Beachem, *Transactions of the American Society of Metals* 56 (1963) 318.
[16] A.R. Rosenfield, *Metallurgical Reviews* 13, review 121 (1968) 29–40 and *Metals and Materials* 2 (1968).
[17] F.A. McClintock, *Journal of Applied Mechanics* 35 (1968) 363–371.
[18] J.R. Rice and D.M. Tracey, *Journal of the Mechanics and Physics of Solids* 17 (1969) 201.
[19] J.R. Rice, in *Mechanics of Fracture,* ASME-AMD 19 (1976) 23–53.
[20] E.B. Glennie, *Journal of the Mechanics and Physics of Solids* 20 (1972) 415–429.

[21] K. Ravi-Chandar and W.G. Knauss, *International Journal of Fracture* 25 (1984) 247–262; 26 (1984) 65–80, 141–154, 193–204.
[22] T. Kanazawa, S. Machida, T. Teramoto and H. Yoshinari, *Experimental Mechanics* (1981) 78–88.
[23] Z. Bilek, in *Crack Arrest Methodology and Applications,* ASTM-STP 711 (1980) 240–247.
[24] A.J. Carlsson, *Transactions of the Royal Institute of Technology,* Stockholm, 189 (1962).
[25] B. Brickstad and F. Nilsson, *International Journal of Fracture* 16 (1980) 71–84.
[26] Z. Bilek, *Scripta Metallurgica* 12 (1978) 1101–1106.
[27] G.T. Hahn et al., Report to the U.S. Nuclear Regulatory Commission, Battelle Columbus Laboratories (1974–1976).
[28] G.C. Angelino, in *Fast Fracture and Crack Arrest,* ASTM-STP 627 (1978) 392–407.
[29] W.G. Knauss, Proceedings of the 6th International Conference on Fracture, New Delhi, Dec. 4–10, 1984.
[30] W.G. Knauss and K. Ravi-Chandar, *International Journal of Fracture* 27 (1985).
[31] J.W. Dally, *Optical Methods in Mechanics of Solids.* Edited by A. Lagarde, Sijthoff and Noordhoff (1980) 692.
[32] A.S. Kobayashi, *Experimental Mechanics* 10 (1970) 106.
[33] A.S. Kobayashi, in *Fracture Mechanics.* Edited by N. Perrone et al. University Press, Charlottesville, Virginia (1978) 481–496.
[34] J. Beinert and J.F. Kalthoff, in *Mechanics of Fracture,* Vol. VII. Edited by G. Sih, Sijthoff and Noordhoff (1981) 281–330.
[35] W.G. Knauss, R. Pfaff and C. Schultheisz. Private Communication. Caltech (1984).
[36] K.S. Kim, *Journal of Applied Mechanics* (1985) to appear.
[37] T. Kobayashi and J.W. Dally, *Crack Arrest Methodology and Applications,* ASTM-STP 711 (1980) 189–210.
[38] P. Manogg, Ph.D. Thesis, University of Freiburg (1964).
[39] P.S. Theocaris and E.E. Gdoutos, *Journal of Applied Mechanics* 39 (1972) 91.
[40] A.J. Rosakis, *Engineering Fracture Mechanics* 13 (1980) 331–347.
[41] A.J. Rosakis and L.B. Freund, *Journal of Applied Mechanics* 48 (1981) 302–308.
[42] A.J. Rosakis and A.T. Zehnder, *Journal of Elasticity* (1985) to appear.
[43] A.J. Rosakis and A.T. Zehnder, work in progress. Caltech (1984).
[44] A.J. Rosakis and K. Ravi-Chandar, GALCIT Report SM 84-2 (March 1984).
[45] W. Yang and L.B. Freund, Brown University Report (1984).
[46] N. Levy, P.V. Marcal and J.R. Rice, *Nuclear Engineering and Design* 17 (1971) 64–75.
[47] A.J. Rosakis and L.B. Freund, *Journal of Engineering Materials and Technology* 104 (1982) 115–120.
[48] R.J. Weimer and H.C. Rogers, *Fast Fracture and Crack Arrest,* ASTM-STP 627 (1977) 359–371.
[49] J. Congleton and B.K. Denton, in *Fast Fracture and Crack Arrest,* ASTM-STP 627 (1977) 336–358.
[50] A.S. Kobayashi et al., *Journal of Engineering Materials Technology* 100 (1978) 402–410.
[51] L. Hodulak, A.S. Kobayashi and A.F. Emery, *Fracture Mechanics: Twelfth Conference,* ASTM-STP 700 (1980) 174–188.
[52] T. Nishioka and S.N. Atluri, *Engineering Fracture Mechanics* 16 (1982) 157–175.
[53] J.F. Kalthoff et al., in *Crack Arrest Methodology and Applications,* ASTM-STP 711 (1980) 109–127.
[54] L. Dahlberg, F. Nilsson and B. Brickstad, in *Crack Arrest Methodology and Applications,* ASTM-STP 711 (1980) 89–108.

Résumé

On présente et on discute certains aspects fondamentaux de la croissance dynamique d'une fissure dans des aciers de construction. La discussion prend la forme d'une comparaison directe des résultats expérimentaux à l'analyse élasto-plastique et tente de clarifier le rôle de l'inertie et de la plasticité du matériau dans le processus de croissance dynamique d'une fissure. On considère en outre la relation qui les critères de la croissance d'une fissure aux modèles micro-mécaniques de croissance des lacunes.

On présente les problèmes potentiels que peuvent surgir dans l'analyse des données obtenues par des mesures directes optiques ou par des simulations numériques de la croissance d'une fissure. On démontre que des erreurs importantes dans les enregistrements de vitesse causées par des effets d'onde de contrainte sont la source principale d'incertitudes dans l'interprétation des résultats expérimentaux.

International Journal of Fracture 27 (1985) 187–201.

Mechanics of crack curving and branching – a dynamic fracture analysis

M. RAMULU and A.S. KOBAYASHI

Department of Mechanical Engineering, University of Washington, Seattle, WA 98195, USA

(Received September 1, 1984; in revised form November 8, 1984)

Abstract

A comparative study on crack curving and branching criteria in dynamic fracture mechanics shows that the criteria based on "advanced cracking" concept correlated best with available experimental data. The crack branching criterion requires as a necessary condition, a critical dynamic stress intensity factor, K_{Ib}, and a sufficient condition involving the crack curving criterion. The criteria are used to predict crack curving and crack branching in dynamic photoelastic experiments involving Homalite-100 and polycarbonate fracture specimens, as well as bursting steel and aluminum pipes.

1. Introduction

In a brittle material, a propagating crack will often depart from its original straight trajectory and curve or split into two or more branches or under high state of stress, divide into a river delta crack pattern. This fragmentation of highly loaded, brittle materials is often a succession of multiple branchings of what was initially a single crack. Typical crack curving and crack branching in a glassy polymer, Homalite-100, are shown in Fig. 1.

Despite the number of comprehensive review articles in the general area of dynamic fracture [1–8], elastodynamic analyses of crack curving and branching are few. An approximate analytical solution for a static curved crack with nonvanishing mode II stress intensity factors [9,10], however, is available. Also, dynamic crack branching has been modeled by an elastostatic solution which was used to predict the crack branching angle. An excellent review of static crack branching, and the inaccuracies and inconsistencies involved in such analyses has been provided by Lo [11]. Pärletun [12] and Kalthoff [13] solved numerically, static crack branching problems with a vanishing mode II stress intensity factor.

2. Dynamic crack curving and crack branching

2.1. Crack curving

Experimental observations of crack curving in brittle materials [14–20] show that the curved crack path differs with specimens and loading conditions. For example, the crack path in a bend specimen under impact loading may have a characteristic S-shape as the crack approaches the compression side of the specimen [14,15] and a double cantilever beam specimen under wedge loading often yields a gradual curving crack path after fracture initiation [16–19]. Slightly deflected crack paths in single edge notch specimens under uniaxial loading have also been reported in [20].

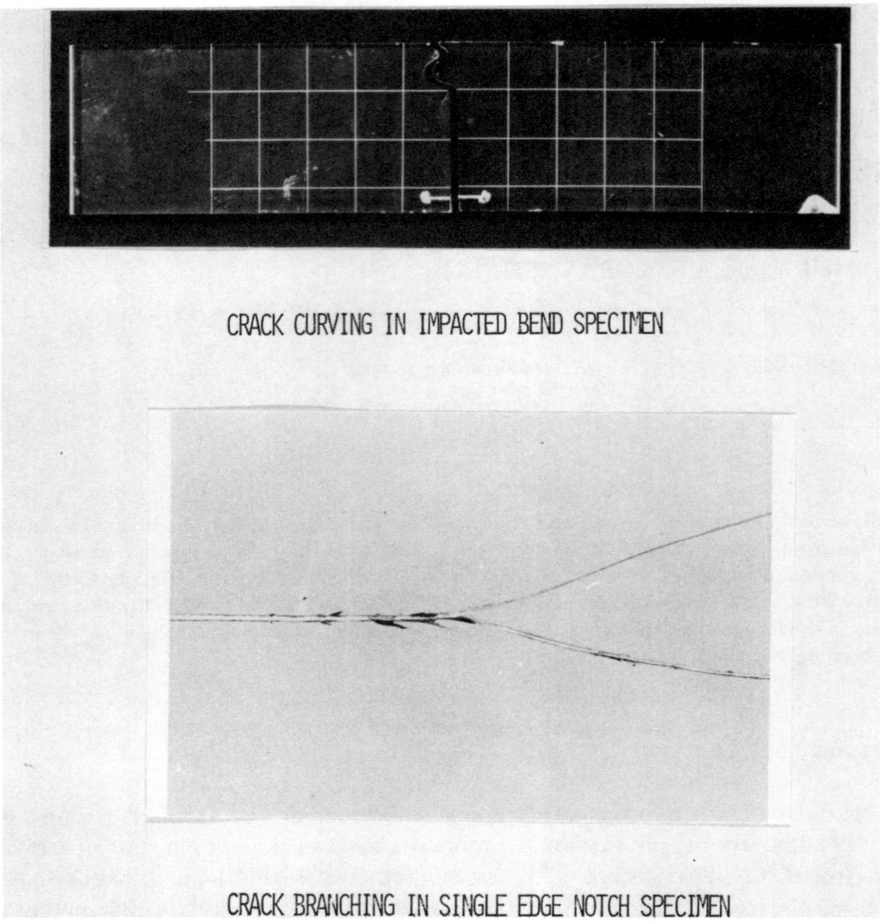

Figure 1. Crack curving and branching.

Directional stability of a straight crack can be enforced either by side grooving or by applying lengthwise compression [17]. Kibler and Roberts [21] showed experimentally, that an increase in the biaxial load increased the apparent fracture toughness and thereby decreased the directional control. Cotterell [18] and A.S. Kobayashi et al. [16] studied the biaxial stress ratio, σ_{xx}/σ_{yy}, at the crack tip and concluded that for straight crack growth, this biaxial stress ratio should be less than unity. Using the same approach, Radon et al. [22] modeled an S-shaped crack as two equivalent-angled cracks or two opposing circular-arc cracks with known modes I and II stress intensity factors, K_{I} and K_{II}, and predicted the curving angles when $K_{\mathrm{II}} = 0$ for varying biaxial stress ratios. In the early work of Benbow and Roesler [17] in 1957, directional stability in square plates was achieved by restricting the movement of the crack faces to parallel outward motion. In DCB specimens, A.S. Kobayashi et al. [16], and Finnie and Saith [23] demonstrated the utility of the lengthwise compression, which is applied parallel to the crack, for crack path stability. More recently, Streit and Finnie [19] developed a directional stability criterion by incorporating the non-singular stress term, σ_{Ox}, which is often referred to as the remote stress component. Directional stability of an extending crack was achieved when a geometrical radius r_0 exceeded or equalled a critical material property r_c. In a series of papers Cotterell [18,24,25], used the crack tip field equations of Williams [26], to show the influence of the nonsingular stress, σ_{0x}, on the directional stability of a propagating crack.

Cotterell concluded that in the absence of K_{II}, the fracture path is stable when $\sigma_{0x} < 0$ and unstable when $\sigma_{0x} > 0$. These conditions, however, were violated in some experiments [20,27,28]. Cotterell and Rice [10] also derived analytically a first order approximation of the necessary conditions for crack curving where the stability of the crack path is governed by the sign of σ_{0x} when $K_{II} = 0$. More recently, Karihaloo et al. [29] derived a curving criterion with non-vanishing dK_{II}/da, when σ_{0x} is zero.

In a recent experimental investigation, Ravi-Chandar and Knauss showed [27,28] that the running crack will have an oscillatory crack path of continuous crack curvings due to the interactions of low amplitude stress waves. In their test specimens, the reflected stress waves from the bottom and top boundaries of the cracked plate impacted the propagating crack tip at different times, changed the crack tip stress field, and induced a change in the crack direction. This experimental evidence suggested that crack curving depends strongly on the specimen geometry and on the loading conditions. Recently, Ramulu and Kobayashi [30] analyzed a series of photoelastic experiments and concluded that crack curving is associated with the stress field acting parallel to the crack tip. This parallel stress appeared to contribute to the crack curving observed by Cotterell [24] and Finnie et al. [19,23]. The latter analyses were elastostatic in nature and no theoretical analyses of dynamic crack curving is available to date. Ravi-Chandar and Knauss [27,28], and Ramulu and A.S. Kobayashi [30], however, characterized dynamic crack curving experimentally, as will be discussed later in this paper.

2.2. Crack branching

Crack branching has been frequently observed during the past 18 years of dynamic fracture research at the University of Washington [31], the University of Maryland [32], California Institute of Technology [27], in Europe [13,33–37], and most recently at the University of Rhode Island [38]. Research on crack branching seeks the criterion which controls the onset of bifurcation or crack branching. Among the possibilities considered were the distortion of the crack tip stress field at a critical velocity and the critical stress intensity factor and strain energy release rate at the onset of branching. The size of the fracture mirror zone and the branching event have been correlated and the fracture surface topography and surface roughness have been studied [27,28,31,39]. An increase in the fracture surface roughness prior to branching was observed consistently in all reported investigations.

2.2.1. Critical velocity criterion

Long before solutions to elastodynamic crack problems were available, Yoffe [40] presented the steady-state solution for a crack of constant length, moving along a straight line in an infinite two-dimensional medium under remote tractions. Yoffe's pre-branching analysis showed that the maximum circumferential stress, $\sigma_{\theta\theta}$, exhibits two symmetrical maxima along the crack axis at a crack velocity of about $C/C_1 = 0.33$, where C and C_1 are the crack and dilatational wave velocities, respectively. This critical velocity, which shifted the maximum $\sigma_{\theta\theta}$ orientation away from $\theta = 0$, was suggested as promoting crack branching. Later, Craggs [41] showed through the solution of a semi-infinite crack propagating at a constant velocity, that the critical velocity for crack branching should be about $C/C_1 = 0.40$, and reaffirmed Yoffe's criterion on a critical crack velocity. The solution presented by Baker [42] for a propagating crack, which suddenly appears in an infinite plate, showed that $\sigma_{\theta\theta}$ stress is not a principal stress; Baker suggested that the crack should propagate perpendicular to the direction of the maximum principal stress. The physical implications of this maximum principal stress criterion will be discussed later. Validation of the above theoretical solutions, however, was stymied by the lack of detailed experimental data on the dynamic state of stress in the vicinity of a moving crack tip.

Experimentally observed crack velocities at crack branching are much smaller than the theoretically predicted critical crack velocities for branching. Bowden et al. [43] showed experimentally, that a crack need not propagate at a constant terminal speed before branching occurs, but may accelerate up to the instant of branching, which is then accompanied by a drop of 5 to 10 percent in the crack speed after branching. Döll's [44] experimental observations on plate glass and FK-52 glass indicated that the constant branching velocity of $C/C_1 = 0.28$ and 0.3 was lower than the theoretical value of $C/C_1 = 0.40$. The precise ultrasonic ripple marking technique used by Kerkhoff [36] also showed that the crack velocity decreased about 10 percent in glass immediately after branching, whereas Schardin [45] and Acloque [46] observed no change in crack velocities during branching in plate glass and a 6 percent change in pre-stressed glass, respectively. Crack branching velocities in various steels reported by Carlsson [47–49], Irwin [50], Hahn et al. [51], Congelton et al. [33–35], and Weimer and Rogers [52,53] were less than $C/C_1 = 0.25$. A.S. Kobayashi et al. [31,54,55], and Dally et al. [8,32,56] studied crack branching using dynamic photoelasticity. Ravi-Chandar and Knauss [27,57] used the method of caustics in a glassy polymer, Homalite-100 specimens and observed that crack branching occurred consistently at velocities well below the crack velocity ratio of $C/C_1 = 0.25$. Similar observations were made by Paxson [58] and Doyle [59]. The observed low crack branching velocities, which hardly decreased after branching, showed that a postulated critical crack velocity could not be a prerequisite to crack branching in these materials.

2.2.2. Critical stress intensity criterion

To date, published attempts for analyzing crack branching sought the necessary condition for branching by comparing the stress states prior to and after branching [60,61]. Since crack branching was observed at lower velocities, criticality of other fracture parameters at subcritical lack velocities must be considered [62]. Such parameters could be a critical stress intensity factor or a critical stress in a region ahead of the crack tip. The latter would facilitate development of secondary cracks at an inclination to the propagation direction of the primary crack.

Clark and Irwin [62] concluded that crack branching occurs by advanced cracking, which requires a critical branching stress intensity factor, K_{Ib}. They also stated that the crack velocity approached a limiting velocity before branching. Independently, Congleton et al. [33–35,63] proposed a critical stress intensity factor criterion for crack branching based on advanced cracking. A model of crack branching, which is based on advanced cracking of a Griffith crack, of length 2a, located at a distance, r_c, ahead of the propagating crack formed the micro-mechanic basis of their criterion. Further improvement of this crack branching model was made by Congleton [63], who postulated an advance crack of a penny shape [64]. He then derived a relation between the branching stress intensity factor, K_{Ib}, and the fracture toughness, K_{Ic}. Based on velocity arguments, Congleton suggested that when $n \geqslant 2$, unloading of the main crack is prevented, where n is a constant involved in the assumed relation, $r_c = na$. The critical distance, r_c, could not be determined and thus, $r_c \geqslant 2a$ was assumed. The final linking of the off-axis micro-cracks with the primary crack in the Congleton model resulted in crack branching, and is similar to that assumed by Carlsson [48]. Dally [8], restating the Congleton model, argued that $n = 2$ is somewhat arbitrary and suggested that the "n" be an open parameter which characterizes the size of the fracture process zone at the instant of branching.

The fracture surface topography of mirror, mist, and hackle zone, is clearly visible in ceramics, brittle polymers, and on a larger scale, in rocks. Hence, the fracture mirror zones

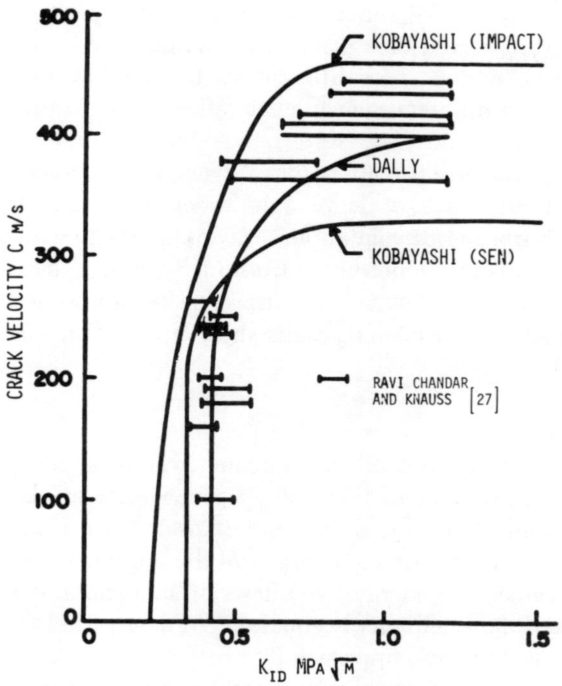

Figure 2. K_{ID} vs. C relations, Homalite-100.

have been utilized to study the strength of ceramics by correlating the algebraic product, A, formed with the mirror radius, $\sqrt{r_m}$, and the fracture stress, σ_f, where A is commonly referred to as the mirror constant and frequently treated as a material constant [65]. The dimensions of the constant are similar to the fracture toughness and thus attempts were made to relate it to the crack branching stress intensity factor, K_{Ib}. The mirror constant, A, has been used in several practical applications, such as to estimate the fracture surface energy [65–67], to determine the magnitudes of the residual stress [67–68], and to predict the flaw size [69–71] in ceramics. However, experimental results obtained by Abdel-Latif et al. [72] and Kishimoto et al. [73] showed that A or K_{Ib} is not a constant but depends on loading conditions. The use of a mirror constant, A, as a branching criterion has two drawbacks: (1) the specimen size does not enter into the calculation of A; and (2) a two-dimensional flaw-shape correction was used. The shape of the mirror, which is a surface flaw, is hard to evaluate because of its three-dimensional shape [73]. The mechanisms of mist, hackle, and crack branching in ceramics have been reviewed extensively by Rice [74].

Available experimental data on crack branching varied from material to material and also appeared to be a function of the test method. Experimental evidence by A.S. Kobayashi et al. [31,75] showed that crack branching consistently occurred in Homalite-100, singled edge notch (SEN) specimens only when K_{I} reached its maximum value with a negligible K_{II}. This study also showed that the branching stress intensity factor is not a unique material property in agreement with Kishimito et al. [73] but contradicts the results of Congleton [63] and Döll [44]. Recently, Rossmanith [76] indicated that increasing crack tip bluntness is the contributing factor in enhancing the stress intensity factor which leads to branching. Dally et al. [8,32,56], showed the existence of a Γ-shaped plateau, which was considered a necessary condition for branching in the $C - K_{\mathrm{ID}}$ relation, where K_{ID} is the dynamic fracture toughness. The experimental studies of dynamic crack propagation by

Ravi-Chandar and Knauss [77,78], indeed showed that a unique $C - K_{ID}$ relation did not exist, and that K_{ID} varied even though the crack velocity remained constant. Relations between crack velocity C and K_{ID} for different fracture test specimens of Fig. 2 show the "plateau", as well as the differences in plateau values which varied with specimen and loading conditions.

The stress wave hypothesis [48,79,80] requires strong interactions between the propagating crack tip and stress waves of large amplitudes. Carlsson [48] hypothesized that micro-cracks form when a tensile dilatational wave impacts the crack front and that the main crack could fork only in the presence of residual stresses. The stress wave hypothesis for crack branching is thus a plausible cause but is not a necessary condition for branching since available experimental results show crack branching without stress wave interaction [27,28,55].

2.2.3. The energy criterion

The energy rate criterion requires crack branching when the strain energy release rate exceeds the energy which is dissipated by a single propagating crack. Based on experimental studies, Johnson and Holloway [81,82] and Bansal [83] attributed branching to the excessive supply rate of the driving energy to the propagating crack tip. Recently, Rabinovitch [84] discussed the energy hypothesis of Johnson and Holloway [82] and of Bansal [83] and showed that their energy equations were not valid for a crack propagating at the limiting velocity. Eshelby's approach [85] to crack branching is also based on the energy balance at the onset of branching at a crack velocity larger than $0.3C_1$.

By examining the experimental results for different materials with identical specimen geometry, Rossmanith and Irwin [86] and Rossmanith [76] indicated that heat loss had to be accounted for in predicting crack branching. Fuller et al. [87] indeed observed the increase in temperature at the onset of crack branching.

2.3. Crack branching angle

Theories used to predict crack branching angles under static loading conditions are the maximum circumferential stress theory [88], the minimum strain energy density factor theory [89], and the maximum strain energy release rate theory [90–93]. Theoretically predicted branching angles based on static analyses are larger than the experimentally observed branching angles. Sih [94] predicted branching angle of 15–18 degrees under a pure mode I dynamic loading condition by using the minimum strain energy density criterion. Since this criterion is sensitive to variations in Poisson's ratios, it could be verified experimentally by examining crack branching data of different materials. By using the maximum stress criterion, Kalthoff [13] found that crack branches with a small branching angle tended to repel one another when $K_{II}/K_I > 0$. Also, branches with a large initial branching angle tended to attract each other when $K_{II}/K_I < 0$. Crack propagation along the original direction was possible when $K_{II} = 0$. The branching angles measured by Kalthoff agreed with his analytical prediction. Elastostatic solutions of Kitagawa et al. [95,96] and Vitek [97] arrived at a similar conclusion, where the crack branched with a branching angle of 18 degrees at zero K_{II} and agreed well with the analytical results of Lo [11].

The experimental results for the branching angle by Kalthoff [13] and those of A.S. Kobayashi [31] and Christie [98] indicate that the crack tip stress state governs the magnitude of the branching angle. Congleton [63] observed crack branching angles of about 40–80 degrees in a bursting steel pipe. Crack branching angles observed by Nakasa and Takei [99], Bullen et al. [100] and others in ceramics [101], all involve positive σ_{0x}, which resulted in larger crack branching angles in these experiments. Indeed, recent experimental results on ceramics [74] showed that the remote tensile biaxiality yielded a larger crack branching angle.

3. Experimental evidences of advanced cracking

Fracture topography and high-speed photography show that fracture is a continuous process of crack initation and crack propagation [8,27,39,48,102–104]. Formation of secondary cracks, which grow with the advancing main crack tip, creates a region of energy dissipation and generates the roughness of the fracture surface. Figure 3 shows the dynamic photoelastic records of a fracturing Homalite-100 specimen just prior to, and at the moment of crack branching, as well as the evolution of multiple cracks which were subsequently generated under a higher crack tip stress field. Some secondary cracks do not link up with the main crack and stop growing after being passed over by the crack tip. Although all materials have an inherent array of microflaws, which grow and coalesce under load, the nucleation phase involving these secondary cracks prior to macro-crack growth is often neglected in the analysis.

The nature of the mirror, mist, and hackle regions on the fracture surface has been investigated as early as 1950 by Kies et al. [105,106] and in the late sixties by Johnson and Hollway [81]. The region of transition, mist, contained a number of geometries resembling parabolas or hyperbolas and semicircular "ribs" in the region of greatest roughness. Such geometries can be associated with an "advance" crack forming parallel to the main fracture plane and resulting in a rough fracture surface [107]. The generation of advance cracks ahead of the main crack was indeed observed by Ravi-Chandar and Knauss [39] in a high-speed photographic record of a fracturing Homalite-100 specimen. From this experimental observation, they concluded that the roughness of the fracture surface results from "independent" fracture origins, which generate the many parabolic and hyperbolic markings, at the front of the main crack. This experimental evidence of advance crack generation justify the use of advance cracks in modeling crack curving and branching mechanisms. Based on these experimental results, Ravi-Chandar and Knauss [39] proposed a mechanism for crack branching which, in essence, is similar to the authors' approach [30,108,109].

From the above brief review, the following salient observations are obtained:

1. No theoretical analysis on dynamic crack curving and branching is available.
2. The existence of the $C - K_{\mathrm{ID}}$ plateau may be a necessary, but not a sufficient, condition for crack branching.

Figure 3. Incipient crack branching photoelastic record of Homalite-100 specimen.

Second frame, 37μ seconds Third frame, 71μ seconds Fourth frame, 103μ seconds

3. The branching stress intensity factor, K_{Ib}, could be a material property.
4. The crack branching angle, θ_c, could depend on the specimen geometry and loading.

4. Near field state of stresses of a rapidly running crack

The crack tip dynamic state of stress under mixed mode conditions was given by Freund and Clifton [110] and the general solution of a running crack was reported by Nishioka and Atluri [111] in terms of the local rectangular, (x, y), and polar coordinates (r, θ). Under pure mode I loading conditions (i.e., $K_{II} = 0$), the dynamic singular crack tip stress field for small θ values differs with the corresponding static stress field in that the largest (tensile) principal, singular stress component acts parallel to the x-axis not only in mode I loading but also in mixed mode loading for smaller values of K_{II}/K_I [30]. Furthermore, the biaxial stress ratio of σ_{xx}/σ_{yy} is roughly equal to unity for both static and dynamic loads up to $C/C_1 < 0.10$. The σ_{0x} term exists only in the σ_{xx} stress component of mode I loading, and is zero in all other components of mode I and mode II stress fields. Situations arise however, for which a single parameter, K_I, characterization of the crack tip stress field is not adequate due to the large spreading of the fracture process zone or the reduction in the size of the K-dominated singular zone [8]. Even in a brittle material, roughening of the fracture surface due to the spreading of the advance cracks and incipient branching can substantially enlarge the fracture process zone. In such cases, it is desirable to incorporate additional stress field parameters in order to characterize the stresses with those associated with a straight and sharp crack. This inevitable involvement of the higher order terms, forms the basis of the following crack curving and branching analyses of a moving crack.

The higher order terms in the crack-tip stress field not only initiate crack curving or branching but also influence their directions. The singular principal stress, σ_1, attains its maximum value at 60 degree with respect to the crack when the crack velocity $C/C_1 \rightarrow 0$, and shifts to 60-105 degrees with increasing crack velocity. The angular position of the maximum stress, unfortunately does not correlate with the experimentally observed crack curving or branching angles when additional higher order terms are introduced in computing the maximum stress. These poor correlations between the experimental and theoretical crack branching angles compel one to look at an alternative, the maximum circumferential stress, which is maximum at $\theta = 0$ and almost constant within $\theta = \pm 15$ degrees. This maximum circumferential stress will generate advance cracks when $C/C_1 < 0.32$ and will enhance openings of the off-axis micro-cracks which are located at a critical distance off the propagating crack tip.

5. Micro-mechanics of crack curving and crack branching

The hypothetical crack curving and crack branching mechanisms based on crack-tip micro-cracking is explained in the following [109]. First the micro-cracks nucleate from the inclusions and voids in the vicinity of the crack tip and grow as a result of the imposed crack-tip stress field. Continual crack growth results in coalescence of the micro-cracks in the crack-tip region and reduces the stress intensity factor by the increase in the compliance of the crack-tip core region. Simultaneously, the remote stress component, σ_{0x}, which acts parallel to the crack and which is the second order term in the crack-tip stress field, increases in magnitude [4,112]. The increased magnitude in the stress parallel to the crack in turn activates the microcracks away from the crack-tip plane. If the micro-cracks are sufficiently close to the main crack tip, they will tend to divert the crack away from its original plane [4,112–114]. A single crack-tip diversion results in crack curving and multiple crack-tip diversion results in crack branching. Also, with an increasing stress intensity factor, the micro-cracks grow sufficiently large to engulf the main crack tip which now depicts a blunt crack. The larger energy release rate associated with the apparent blunt crack tip will in turn require the generation of branched cracks for larger energy dissipation. As a result, the main crack will extend through this damaged region and leave in its wake a series of distinct attempted branches [114]. At a sufficiently high applied stress intensity factor of K_I, the branched cracks will continue to propagate and thus complete successful crack branches [63]. In the foregoing cases, the necessary prerequisite for crack curving and branching is the growth of micro-cracks into macro-cracks with eventual hookup to the main crack. As shown in previous papers [19,30,108,109,115–117], the remote stress component, σ_{0x}, which is acting parallel to the direction of the crack propagation, controls the location and orientation of micro-cracking and hence governs the directional stability of an extending crack.

5.1. Dynamic crack curving

With the foregoing assumption that the advanced off-axis micro-cracks dictate the direction of crack propagation, either of the two dynamic crack curving criteria [30] by the authors, namely the maximum circumferential stress or the minimum strain energy density criterion, can be used to predict crack curving. Both criteria predict nearly identical crack curving angles in the experimentally observed crack velocity range and in the presence of "smaller" values of K_{II}. The angular position of the micro-cracks with respect to the crack tip, depends both on the micro-structure of the material and on the stresses at a characteristic distance, r_0, from the crack tip. These two parameters are linked together in a criterion involving the directional stability of an extending crack [30] and is a dynamic extension of Streit and Finnie's directional stability criterion [19] for quasi-statically extending crack.

The onset of crack curving of a rapidly propagating crack is governed by the stability of the propagating straight crack and is assumed to occur when $r_0 \leqslant r_c$. This r_c is a characteristic distance derived from the above-mentioned directional stability criterion which suddenly becomes unstable for $\sigma_{0x} > 0$ or in the presence of a smaller value of K_{II}. The corresponding angle, θ_c, for a maximum $\sigma_{\theta\theta}$, under a pure mode I dynamic crack extension can be determined through a transcendental relation involving the state of stress and r_c. The r_c associated with this directional stability model is assumed to be a critical material property and the crack will curve at the angle θ_c [30].

5.2. Dynamic crack branching criterion

The aforementioned micro-mechanics involved in crack curving and crack branching indicates that the crack branching criterion requires both a critical stress intensity factor that is accompanied by the characteristic distance $r_0 \leqslant r_c$. The crack branching criterion can thus be stated as:

$K_I \geqslant K_{Ib}$ Necessary condition

$r_0 \leqslant r_c$ Sufficient condition.

6. Recent experimental results

6.1. Dynamic crack curving in photoelastic specimens

The validity of the above crack curving criterion was verified by results involving photoelastic measurements on Homalite-100 specimens. Crack curving was consistently observed in all of the above specimens when $r_0 \leqslant r_c$ and yielded an average value $r_c = 1.3$ mm at the onset of curving. A small K_{II} co-existed immediately after crack instability, possibly due to advance cracks, and changed the direction of crack propagation. Experimental details of this series of tests can be found in [14,16,20,30].

Further verification of this criterion was made by Sun et al. [115] with dynamic photoelastic experiments involving polycarbonate, double-edge crack tension specimens. Crack configurations involved were either offset parallel cracks, offset slanted cracks, and symmetrically located twin cracks which were used in this series of dynamic photoelastic study. The annealed thin polycarbonate specimens with blunt starter cracks exhibited brittle fracture with shear lips less than 10 percent of the thickness. At the onset of crack curving, the average r_c value was 0.5 mm which is consistent with the r_c estimated by Theocaris [118]. The theoretically predicted crack curving angles, either in pure mode I or modes I and II loading conditions, in the presence of σ_{0x}, were within one degree of the corresponding measured values.

6.2. Dynamic crack branching in photoelastic specimens

Crack branching criteria [108,109] require a critical stress intensity factor to trigger crack branching and a curving criterion to predict the crack branching angle. A review of all

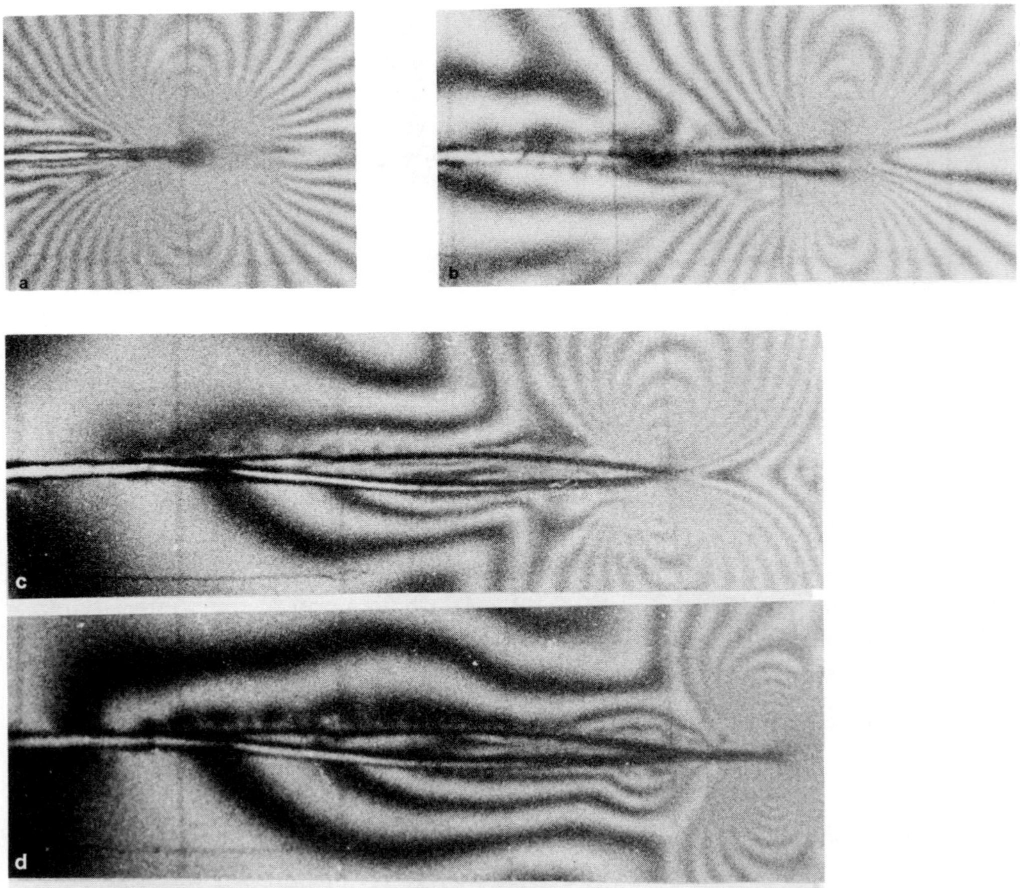

Figure 4. Typical dynamic-photoelastic record of crack branching in a polycarbonate single-edged-notch specimen. Specimen NO-RB-811003-01, (a) Eight frame, 83μ seconds. (b) Eleventh frame, 122 μ seconds. (c) Thirteenth frame, 151μ seconds. (d) Fourteenth frame, 174μ seconds.

past and recent crack branching data on Homalite-100 and polycarbonate photoelastic polymers involving SEN specimens showed that most cracks would bifurcate upon reaching a branching stress intensity factor.

The crack branching results of the Homalite-100 experiments [108] yielded an average r_c of 1.3 mm at branching and a $K_{Ib} = 2.04$ MPa$\sqrt{m}$. At the onset of crack branching, the crack velocities were found to be about 0.18 C_1. This K_{Ib}, which is approximately 4.9 times the fracture toughness of Homalite-100, is in agreement with the measured K_{Ib} of Dally [8].

Figure 4 shows five frames out of a 16-frame dynamic photoelastic record in a 3.2 mm thick, 127 × 127 mm polycarbonate SEN specimen with a blunt starter crack. No attempt was made to analyse these post-branched cracks, but at the onset of branching, $K_{Ib} = 3.2$ MPA$\sqrt{m}$ was obtained. It is interesting to observe in this experiment that the branching angle was 8 degrees and is less than the usually observed angles in the SEN specimens. After branching, post-branch cracks ran parallel and close to each other. These parallel cracks could have propagated together only if they were traveling at the same velocity. The slower of the two, however, arrested, as shown in [108]. The stress intensity factors of these two closely spaced cracks could not be determined since the interaction effects of the two cracks have not been incorporated into the data reduction process. The crack velocity at the onset of branching was found to be about 0.22 C_1 and is consistent with the other five crack branching experiments reported earlier [116].

In the SEN specimens, the energetic requirement of multiple cracking was satisfied but the crack curving criterion could not be satisfied since $\sigma_{0x} < 0$, $K_{II} = 0$. Under such a condition, $\sqrt{r_c}\,\sigma_{\theta\theta}/K_I$ is maximum at $\theta = 0$ for the crack velocity of $C/C_1 < 0.32$. However, the reduction in the angular stress distribution of $\sqrt{r_c}\,\sigma_{\theta\theta}/K_I$ at $\theta = \pm(10\text{-}14)$ degrees is a mere $1.7 - 3.5$ percent which allows the crack to bifurcate and double its energy release rate. Therefore, the predicted crack branching angle is at best an estimated angle [116].

The above-mentioned crack curving as well as the crack branching criteria were also verified in wedge loaded, rectangular double cantilever beam (WL-RDCB), Homalite-100, specimens [108,117], and in pressurized metal pipes of steel [119] and aluminum [120]. Predicted and measured branch angles in wedge loaded, rectangular double cantilever beam (WL-RDCB) specimens and bursting pipes were found to be in good agreement [108,109].

7. Discussions

High-speed photoelastic investigations [14,16] of crack curving reveal that this phenomenon is often associated with the backward tilt of the crack tip isochromatics. This tilt indicates positive values of remote stress, σ_{0x}, observed in dynamic tear test (DTT) specimens under impact loading and double cantilever beam (DCB) specimens under wedge loading. The proposed crack curving criterion predicts the curving angle well. Stress wave interaction and the resulting mixed mode effect also control the crack direction.

The direction of maximum $\sigma_{\theta\theta}$ at a critical radial distance is always at $\theta = 0$ for $\sigma_{0x} < 0$ and $C/C_1 < 0.32$. However, crack branching occurred at an average included crack branching angle of $2\theta_c = 28$ degrees when $C/C_1 < 0.32$ and when $\sigma_{0x} < 0$ in SEN specimens. The differences between the theoretical and experimental results can be explained by examining the instantaneous circumferential stress associated with the onset of crack branching at the critical radial distance. As discussed earlier, the reduction in circumferential stress at $\theta = \pm 14$ degrees is less than 3 percent for the observed crack velocity. Presumably, the propagating crack can extend in any direction within $\theta = \pm 14$ degrees in order to bifurcate and double its energy release rate. The averaged value of the

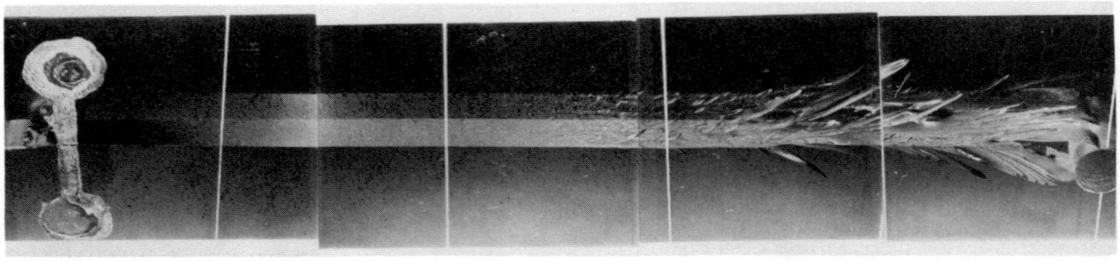

Figure 5. Typical fracture surface in Homalite-100 crack branching experiments.

estimated half-branching angle of 15.4 degrees in polycarbonate [116] and 14 degrees in Homalite-100 specimens is in good agreement with the measured half-branching angle. This agreement demonstrates that the crack branching criterion remains valid despite the predicted self-similar crack extension for $K_{ID} < K_{Ib}$. The calculated one-half branching angle with $\sigma_{0x} < 0$ is the best estimated angle under a pure mode I condition. The dynamic stress intensity factor at the onset of crack branching reached an average maximum value of 2.0 $MPa\sqrt{m}$ in Homalite-100, and 3.3 $MPa\sqrt{m}$ in polycarbonate fracture specimens. The branching stress intensity factors were found to be material-dependent.

The photoelastic patterns of running post-branched cracks in all of our experimental observations initially showed the effect of mixed mode and higher order terms. These results suggest that the crack will run parallel to the compressive stress direction and that the mixed mode isochromatics after branching appeared to be generated by the advance cracks. Therefore, postbranching crack propagation is dependent on the K_{II}/K_I ratio but also on σ_{0x} as well. The crack curving of Fig. 4 shows that the crack curving angle of both parallel running cracks gradually decreased so that the cracks curved toward each other and continued to propagate when the crack branching angle was 8 degrees. Although inconclusive, it appears from the available experimental results, that successful branching is possible only if 2 $\theta_c > 20$ degrees. For $2\theta_c < 20$ degrees, however, branched cracks merged with the main crack and propagated in accordance with Kalthoff's results [13].

Figure 5 shows the typical fracture surface in a 12-mm-thick Homalite-100 SEN specimen associated with crack branching with mirror, mist, and hackle regions. The mist, hackle, and crack branching show the nucleation of advance cracks which resulted in surface roughness [39]. Also, note that the microcracks do not span the entire plate thickness. Some exist on the two plate faces while others occur in the middle of the plate and are continuously curving prior to a successful crack branching. This fracture surface indicates that the σ_{0x} term is compressive and opens the micro-cracks in shear failure.

Work performed so far in dynamic crack curving and crack branching was primarily concerned with two-dimensional problems. In most engineering materials, crack growth, curving, and branching are locally three-dimensional problems associated with advance nucleations of micro-cracks. The high-speed photomicrographic experiments of Ravi-Chandar and Knauss [39,57], illustrate the three-dimensional effects in the nucleation of micro-cracks.

8. Conclusions

The review of the present and of the previous studies on crack curving and crack branching were presented. Criteria based on crack tip micro-mechanics proposed by the authors were used to predict the curving and branching in brittle materials.

Acknowledgement

The work reported here was obtained under ONR Contract NO-0014-76-C-000 NR-064-478. The authors wish to acknowledge the support and encouragement of Dr Y. Rajapakse, ONR, during the course of this investigation.

References

[1] J.D. Achenbach, in *Fracture Mechanics* XII SIAM-AMS Proceedings 12 (1979) 3–20.
[2] L.B. Freund, in *Mechanics of Fracture,* Edited by F. Erdogan, 19 ASME (1976) 105–134.
[3] F. Nilson, *SM Archives* 2 (1977) 205–261.
[4] L.R.F. Rose, *International Journal of Fracture* 12 (1976) 799–813.
[5] J.E. Field, *Contemporary Physics* 12 (1971) 1–31.
[6] M.F. Kanninen, in *Numerical Methods in Fracture Mechanics,* Edited by A.R. Luxmoore and D.R.J. Owen, University College of Swansea, UK (1978) 612–633.
[7] A.S. Kobayashi, in *Fracture Mechanics*, Edited by N. Perrone et al., University Press of Virginia (1978) 481–496.
[8] J.W. Dally, *Experimental Mechanics* 19 (1979) 349–367.
[9] B.L. Karihaloo, L.M. Keer and S. Nemat-Nasser, *Engineering Fracture Mechanics* 13 (1980) 879–888.
[10] B. Cotterell and J.R. Rice, *International Journal of Fracture* 11 (1981) 155–164.
[11] K.K. Lo, *Journal of Applied Mechanics* 45 (1978) 797–802.
[12] L.C. Pärletun, *Engineering Fracture Mechanics* 11 (1979) 343–358.
[13] J.F. Kalthoff, in *Dynamic Crack Propagation,* Edited by G.C. Sih, Noordhoff International Publishing (1973) 449–458.
[14] A.S. Kobayashi and C.F. Chan, *Experimental Mechanics* 16 (1976) 176–181.
[15] J.F. Kalthoff, in *Work Shop on Dynamic Fracture,* Edited by W.G. Knauss et. al., California Institute of Technology (1983) 11–35.
[16] A.S. Kobayashi, S. Mall and M. Lee, *ASTM STP 601* (1976) 274–290.
[17] J.J. Benbow and F.C. Roesler, *Proceedings of the Physics Society* London 70 B (1957) 201–211.
[18] B. Cotterell, *International Journal of Fracture Mechanics* 1 (1965) 96–103.
[19] R. Streit and I. Finnie, *Experimental Mechanics* 20 (1980) 17–23.
[20] W.B. Bradley, "A Photoelastic Investigation of Dynamic Brittle Fracture", Ph.D. thesis, University of Washington (1969).
[21] J.J. Kibler and R. Roberts, *Journal of Engineering for Industry* (1970) 727–734.
[22] J.C. Radon, P.S. Leevers and L.E. Culver, in *Fracture 1972,* 3, University of Waterloo Press (1977) 1113–1118.
[23] I. Finnie and A. Saith, *International Journal of Fracture* 9 (1973) RCR 484–486.
[24] B. Cotterell, *International Journal of Fracture Mechanics* 2 (1966) 526–533.
[25] B. Cotterell, *International Journal of Fracture Mechanics* 6 (1970) 189–192.
[26] M.L. Williams, *Journal of Applied Mechanics* 24 (1957) 109–114.
[27] K. Ravi-Chandar and W.G. Knauss, *International Journal of Fracture* 26 (1984) 141–154.
[28] K. Ravi-Chandar and W.G. Knauss, *International Journal of Fracture* 26 (1984) 189–200.
[29] B.L. Karihaloo, L.M. Keer, S. Nemat-Nasser and A. Oranatnachai, *Journal of Applied Mechanics* 48 (1981) 515–519.
[30] M. Ramulu and A.S. Kobayashi, *Experimental Mechanics* 23 (1983) 1–9.
[31] A.S. Kobayashi, B.G. Wade, W.B. Bradley and S.T. Chiu, *Engineering Fracture Mechanics* 5 (1974) 81–92.
[32] G.R. Irwin, J.W. Dally, T. Kobayashi, W.L. Fourney, M.J. Etheridge and H.P. Rossmanith, *Experimental Mechanics* 19 (1979) 121–128.
[33] J. Congleton and N.J. Petch, *Philosophical Magazine* 16 (1967) 749–760.
[34] S.R. Anthony and J. Congleton, *Metal Science* (1968) 158–160.
[35] S.R. Anthony, J.P. Chubb and J. Congleton, *Philosophical Magazine* 22 (1970) 1201–1216.
[36] F. Kerkhoff, in *Dynamic Crack Propagation,* Edited by G.C. Sih, Noordhoff International Publishing, Leyden (1973) 3–35.
[37] P.S. Theocaris, *Journal of the Mechanics and Physics of Solids* 20 (1972) 265–279.
[38] A. Shukla and S. Anand, "Crack Propagation and Branching Under Biaxial Loading," presented at 17th National Symposium on Fracture Mechanics, Albany, New York (August 1984).
[39] K. Ravi-Chandar and W.G. Knauss, *International Journal of Fracture* 26 (1984) 65–80.
[40] E.H. Yoffe, *Philosophical Magazine* 42 (1951) 739–750.
[41] J.W. Craggs, *Journal of the Mechanics and Physics of Solids* 8 (1960) 66–75.
[42] B.R. Baker, *Journal of Applied Mechanics* 29 (1962) 449–458.
[43] F.P. Bowden, J.H. Brunton, J.E. Field and A.D. Heyes, *Nature* 216 (1967) 38–42.
[44] W. Döll, *International Journal of Fracture* 11 (1975) RCR 184–186.
[45] H. Schardin, in *Fracture*, Edited by B.L. Averbach et al., John Wiley (1959) 297–330.
[46] P. Acloque, *Silicate Industrials* 28 (1963) 323–331.
[47] J. Carlsson, *Transactions of the Royal Institute of Technology, Stockholm* 189 (1962) 2–55.
[48] J. Carlsson, *Transactions of the Royal Institute of Technology, Stockholm* 205 (1963) 3–38.

[49] J. Carlsson, *Transactions of the Royal Institute of Technology, Stockholm* 207 (1963) 3–26.
[50] G.R. Irwin, *ASTM STP 627* (1977) 7–18.
[51] G.T. Hahn, R.G. Hogland and A.R. Rosenfield, in *Fracture 1977*, 2, University of Waterloo Press (1977) 1333–1338.
[52] R.J. Weimer and H.C. Rogers, in *Proceedings of the Second International Conference on Mechanical Behavior of Materials,* Boston (1976) 1473–1477.
[53] R.J. Weimer and H.C. Rogers, *Journal of Applied Physics* 50 (1979) 8025–8030.
[54] A.S. Kobayashi, in *Progress in Experimental Mechanics*, Edited by V.J. Parks (1975) 83–97.
[55] A.S. Kobayashi and M. Ramulu, *Experimental Mechanics* 21 (1981) 41–48.
[56] T. Kobayashi and J.W. Dally, *ASTM STP 627* (1977) 257–273.
[57] K. Ravi-Chandar and W.G. Knauss, in *Workshop on Dynamic Fracture*, California Institute of Technology (February 17-18 1983) 119–128.
[58] T.L. Paxson and R.A. Lukas, in *Dynamic Crack Propagation,* Edited by G.C. Sih, Noordhoff International Publishing, Leyden (1973) 415–426.
[59] M.S. Doyle, *Journal of Material Science* 18 (1983) 687–702.
[60] J.D. Achenbach, in *Prospects of Fracture Mechanics,* Edited by G.C. Sih et al., Noordhoff International Publishing (1974) 319–336.
[61] J.D. Achenbach, *Journal of Elasticity* 9 (1979) 113–129.
[62] A.B.J. Clark and G.R. Irwin, *Experimental Mechanics* 6 (1966) 321–330.
[63] J. Congleton, in *Dynamic Crack Propagation,* Edited by G.C. Sih, Noordhoff Publishing (1973) 427–483.
[64] R.A. Sack, *Proceedings Physics Society* London 58 (1946) 729–736.
[65] J.J. Mecholsky, R.W. Rice and S.W. Freiman, *Journal of American Ceramic Society* 57 (1974) 440–443.
[66] J.J. Mecholsky, S.W. Freiman and R.W. Rice, *ASTM STP* 645 (1978) 363–379.
[67] H.P. Kirchner and R.M. Gruver, *Philosophical Magazine* 27 (1973) 1433–1446.
[68] H. Kirchner and W.J. Kirchner, *Journal of American Ceramic Society* 62 (1979) 198–202.
[69] H.P. Kirchner, R.M. Gruver and R.E. Walker, *Journal of American Ceramic Society* 56 (1973) 17–21.
[70] J.J. Mecholsky, S.W. Freiman and R.W. Rice, *Journal of Material Science* 11 (1976) 1310–1319.
[71] H.P. Kirchner, R.M. Gruver, M.V. Swain and R.C. Garvie, *Journal of American Ceramic Society* 64 (1981) 529–533.
[72] A.I.A. Abdel-Latif, R.C. Bradt and R.E. Tressler, *International Journal of Fracture* 13 (1977) 349–359.
[73] K. Kishimoto, S. Aoki and M. Sakata, *Archives Mechanics* 33 (1981) 947–956.
[74] R.W. Rice, *ASTM STP 827* (1984) 5–103.
[75] A.S. Kobayashi and S. Mall, *Experimental Mechanics* 18 (1978) 11–18.
[76] H.P. Rossmanith, "Crack Branching in Brittle Materials Part I", University of Maryland Photomechanics Laboratory Report (1977–1980).
[77] K. Ravi-Chandar and W.G. Knauss, *International Journal of Fracture* 20 (1982) 209–222.
[78] K. Ravi-Chandar and W.G. Knauss, *International Journal of Fracture* 25 (1984) 247–262.
[79] B.R. Lawn and T.R. Wilshaw, *Fracture in Brittle Solids,* Cambridge University Press (1975) 100–104.
[80] H.P. Rossmanith and A. Shukla, *Experimental Mechanics* 21 (1981) 415–422.
[81] J.W. Johnson and G.W. Holloway, *Philosophical Magazine* 17 (1967) 899–901.
[82] J.W. Johnson and G.W. Holloway, *Philosophical Magazine* 19 (1966) 731–743.
[83] G.K. Bansal, *Philosophical Magazine* 35 (1977) 935–944.
[84] A. Rabinovitch, *Philosophical Magazine* 40 (1979) 873–874.
[85] J.D. Eshelby, in *Inelastic Behavior of Solids*, Edited by Kanninen et al., McGraw-Hill New York (1970) 77–115.
[86] H.P. Rossmanith and G.R. Irwin, "The Analysis of Dynamic Isochromatic Crack Tip Stress Patterns", University of Maryland Report (1979).
[87] K.N.G. Fuller, P.G. Fox, and J.E. Field, *Proceedings of the Royal Society* (London) A-341 (1975) 537–557.
[88] F. Erdogan and G.C. Sih, *Journal of Basic Engineering* 85(D) (1963) 519–527.
[89] G.C. Sih, in *Methods of Analysis and Solutions of Crack Problems* 1, Edited by G.C. Sih, Noordhoff International Publishing, Lyden (1973) 21–45.
[90] K. Palaniswamy and W.G. Knauss, in *Mechanics Today*, 4, Edited by S. Nemat-Nasser, Pergammon Press (1978) 87–148.
[91] M.A. Hussain, S.L. Pu, and J. Underwood, *ASTM STP 560* (1974) 2–28.
[92] C.H. Wu, *Journal of Applied Mechanics* 45 (1978) 553–558.
[93] H. Anderson, *Journal of the Mechanics and Physics of Solids* 17 (1969) 405–417.
[94] G.C. Sih, in *Elastodynamic Crack Problems,* 4, Noordhoff International Publishing, Leyden (1977) XVII-XLVII.
[95] H. Kitagawa, R. Yuuki and T. Ohira, *Engineering Fracture Mechanics* 7 (1975) 515–529.
[96] H. Kitagawa and R. Yuuki, in *Fracture 1977*, Edited by D.M.R. Taplin, 3, University of Waterloo Press (1977) 201–211.

[97] V. Vitek, *International Journal of Fracture Mechanics* 13 (1977) 481–501.
[98] D.G. Christie, *Transactions of the Institute of Glass Technology* 36 (1952) 74–89.
[99] K. Nakasa and H. Takei, *Engineering Fracture Mechanics* 11 (1979) 739–751.
[100] E.P. Bullen, F. Henderson and H. Wain, *Philosophical Magazine* 21 (1970) 689.
[101] R.E. Bullock and J.L. Kaae, *Journal of Material Science* 14 (1979) 920–930.
[102] W.B. Bradley and A.S. Kobayashi, *Experimental Mechanics* 10 (1970) 106–113.
[103] A.S. Kobayashi, S. Mall and W.B. Bradley, *Proceedings of the Twelfth Society of Engineering Science,* University of Texas at Austin (October 1975).
[104] A.S. Kobayashi and M. Ramulu, in *Mixed Mode Crack Propagation,* Edited by G.C. Sih and P.S. Theocaris, Sijthoff & Noordhoff (1981) 163–172.
[105] J.A. Kies, A.M. Sullivan and G.R. Irwin, *Journal of Applied Physics* 21 (1950) 716–720.
[106] I. Wolock, J.A. Kies and S.B. Newman, in *Fracture,* Edited by B.L. Averbach et al., Technology Press and Wiley (1959) 250–264.
[107] G.R. Irwin, in *Handbuch Der Physik* VI, Edited by S. Flugge, Springer-Verlag (1958) 551–590.
[108] M. Ramulu, A.S. Kobayashi and B.S.-J. Kang, *Fracture Mechanics* (15th), ASTM STP 833 (1984) 130–142.
[109] M. Ramulu, A.S. Kobayashi and B.S.-J. Kang, *Journal of Pressure Vessel Technology* 104 (1983) 317–322.
[110] L.B. Freund and R.J. Clifton, *Journal of Elasticity* 4 (1974) 293–299.
[111] T. Nishioka and S.N. Atluri, *Engineering Fracture Mechanics* 17 (1983) 1–22.
[112] J.R. Rice, *Journal of the Mechanics and Physics of Solids* 22 (1974) 17–26.
[113] J.R. Rice and M.A. Johnson, in *Inelastic Behavior of Solids,* McGraw Hill (1970) 641–672.
[114] R.O. Richi, J.F. Knott and J.R. Rice, *Journal of the Mechanics and Physics of Solids* 21 (1973) 395–410.
[115] Y.-J. Sun, M. Ramulu, A.S. Kobayashi and B.S.-J. Kang, in *Developments in Theoretical and Applied Mechanics,* Edited by T.J. Chung and G.R. Karr, The University of Alabama Huntsville (1982) 203–218.
[116] M. Ramulu, A.S. Kobayashi, B.S.-J. Kang, and D.B. Barker, *Experimental Mechanics* 23 (1984) 431–437.
[117] M. Ramulu and A.S. Kobayashi, *Engineering Fracture Mechanics* 18 (1983) 1087–1098.
[118] P.S. Theocaris and G. Papadopoulos, *Journal of Applied Mechanics* 49 (1982) 81–86.
[119] E.A. Almond, N.J. Petch, A.E. Wraith and E.S. Wright, *Journal of Iron and Steel Institute* 207 (1969) 1319–1323.
[120] R.W.E. Shannon and A.A. Wells, in *Proceedings of International Symposium on Pipeline,* Paper No. 17, Newcastle upon Tyne (1974).

Résumé

Une étude comparative sur les critères d'incurvation et d'arborescence d'une fissure en mécanique de rupture dynamique montre qu'un critère basé sur le concept de "fissuration avancée" présente la meilleure corrélation avec les données expérimentales disponibles. Le critère d'arborescence d'une fissure requiert comme condition nécessaire un facteur d'intensité de contrainte dynamique critique, K_{Ib}, et comme condition suffisante un critère tenant compte de l'incurvation de la fissure. Les critères sont utilisés pour prédire l'incurvation et l'arborescence d'une fissure au cours d'expériences photo-élastiques en condition dynamique mettant en oeuvre de l'Homalite 100 et du polycarbonate ainsi que des tubes d'acier et d'aluminium en cours d'explosion.

International Journal of Fracture 27 (1985) 203–213.

Dynamic crack branching in brittle solids

J.P. DEMPSEY * and P. BURGERS **

* *Department of Civil and Environmental Engineering, Clarkson University, Potsdam, NY 13676, USA*
** *Hibbitt, Karlsson and Sorensen, Inc., 35 S. Angell St., Providence, RI 02906, USA*

(Received September 1, 1984; in revised form November 2, 1984)

Abstract

A state-of-the-art review of recent and on-going analytical and numerical research on dynamic crack branching in brittle solids is presented. First, fundamental experimental evidence concerning the physics of the branching mechanism is reviewed. Next, elastostatic work dealing with crack branching under an arbitrary angle and crack branching fracture criteria is summarized. The idealizations inherent in the mathematical models under investigation are emphasized. The analytical and numerical schemes that have recently yielded several much needed dynamic self-similar crack branching solutions are discussed. A perturbation-type approximation to the initial-boundary conditions whereby the artificial constraint of self-similarity is removed and variable velocity crack branching can be treated is discussed also. Finally, new directions and the needs of future research are identified.

1. Introduction

Dynamic crack division occurs most readily in brittle solids. These include glass, rock, and polymers and metals subjected to low temperature environments or high rates of loading. The dynamic fracturing of brittle solids is characterized by high crack speeds and, if crack arrest does not occur, subsequent crack branching, further crack propagation and branching, and so on [1–6]. The micro-structural properties of materials affect the velocity at which branching occurs since in crystals with well defined cleavage planes, or in solids with a preferred orientation, higher crack tip speeds may be required to initiate branching [4,7]. Loading in the far-field and stress wave interaction may have a similar effect [6]. It is well known, for instance, that compression parallel to the plane of crack propagation can be used to control the stability of brittle fracture and delay crack branching (see [6] and [8] for experimental evidence and [9] for analytical results).

From an engineering viewpoint, an increased knowledge of the factors governing crack branching is useful in a multitude of ways. In structures comprised of huge welded steel plates [10] and in the smooth cleavage of expensive gem stones [7], crack branching must be prevented. In rock quarrying and mining or oil and gas recovery operations, crack branching must be controlled either to prevent the production of oversized fragments or to achieve the desired porosity, respectively [11,12]. From a materials testing viewpoint, crack branching complicates the task of evaluating the propagating crack energy release rate, fracture energy, stress intensity, and toughness, all of which characterize dynamic fracture processes [13].

Experimental work on dynamic crack branching began with Schardin [1] who conducted experiments on glass and photographed the evolution of the pattern of bifurcating cracks. Crack branching has since been observed in a variety of materials including commercial and toughened glass plates, pressurized glass, steel, and aluminum pipes, crystalline solids, rock, and brittle polymers such as Homalite-100, Araldite, and PMMA. Contributions to the literature that are representative of the above studies include articles by Ravi-Chandar and Knauss [3–6], Field [7], Hahn et al. [13], Bieniawski [14], Dally [15], Congleton [16], Kobayashi et al. [17–20], Kalthoff [21], and Aoki and Sakata [22]; these publications should also enable the reader to trace earlier contributions.

From the experimental evidence gained to date, it is apparent that the dynamic branching event is inherently three-dimensional. An interesting high speed photomicrographic study of the microstructural aspects of dynamic fracture was performed by Ravi-Chandar and Knauss [4]. By capturing the evolution of the crack branching process, they observed that there is no single instant of crack branch initiation, and in fact that there are multiple branching attempts in the process zone made by the propagating crack. These microbranches do not span the plate thickness, some occurring on the faces of the plate while others are entirely in the middle of the plate. Only a few of the microbranches grow while the rest are arrested. The branches evolve from the microcracks which are initially parallel to the main crack, but deviate smoothly from the original crack orientation.

The possibility that crack branching is the result of many interacting microcracks or microbranches was alluded to by Field [7], Dally [15], and Congleton [16]. The latter author also tried to estimate the stress intensity factor at branching by considering a small Griffith crack placed ahead of the main crack. This appealing and plausible hypothesis has proved very difficult to verify experimentally. However, pursuant to their observations in [4], Ravi-Chandar and Knauss [5] argue definitively that the crack branching mechanism is governed by the nucleation, coalescence and subsequent interaction of advance (off-axis) microcracks.

From a macroscopic viewpoint, the experimental evidence can be summarized as follows. The angle subtended by the new branch – immediately after branching – and the original crack plane typically lies between $\pm 10° - \pm 45°$ for glasses [1–3,22] and brittle polymers [2,3–6,15–19,21]. In the pressurized tube tests on steel [16,20] and aluminum [20], and tests on crystalline solids and other solids with preferred orientations [7], the branching angle can be as high as 90°. The crack velocity immediately after branching generally shows little or no change from the pre-branching velocity, thereby reinforcing the view that branching occurs by a "continuous" mechanism. Far-field loading, in particular compression or tension parallel to the original crack plane, can delay or enhance branching, but not fully suppress it [6]; the branching angles can be similarly influenced. Stress wave interaction can also cause branching [6].

The detailed three-dimensional branching situation is extremely difficult to model. Elastodynamic treatments have great difficulty incorporating any of the microscopic details. The approximation to planar deformation and through-the-thickness straight-front crack propagation is therefore not unreasonable. In brittle fracture, the assumption that small scale yielding applies is also not unreasonable. In other words, the process zone – the region in which the crack rapidly curves – is assumed to be small compared to any characteristic length in the problem. Analytically, the process zone is modelled as a point, giving a well-defined instant of branching. The details of the deformation in the process zone can then presumably be lumped into a fracture criterion as is typically done in linearly elastic fracture mechanics. The determination of this fracture criterion is the ultimate objective.

Much of the analytical work dealing with crack branching under an arbitrary angle with the primary crack is elastostatic in nature, and much of the latter work was motivated by Erdogan and Sih's [23] investigation of the fracture of a straight crack under nonsymmetric loading conditions. For antiplane strain deformations, the elastostatic solutions for a kinked and bifurcated crack were presented by Sih [24] and Smith [25], respectively. A representative selection of contributions to the literature that treat elastostatic in-plane problems involving asymmetrically or symmetrically branched cracks includes studies by Hussain, Pu and Underwood [26], Palaniswamy and Knauss [27], Wu [28], Lo [29] and Hayashi and Nemat-Nasser [30]. The multiply branched crack has been examined using the finite element method by Wilson and Cherepko [31].

The slow growth of cracks in non-uniform stress fields, without any a priori assumption regarding the nature of crack motion, has been treated using perturbation procedures by Banichuk [32], Cotterell and Rice [9] and Karihaloo et al. [33]. The latter studies continue to generate much interest, as shown by recent contributions [34,35]. The path of fracture may be kinked or curved, and differs according to which fracture criterion is examined [9,33–37]. Such studies are mathematically complex because the predictions of the critical stress intensity and branching angle must be based on changes in the crack-tip stress-strain field that are caused by branching of the main crack. There have been proposals to determine the conditions for crack branching from the stress-strain field around the unbranched crack, using the *prior* stress field criteria [36,37] which are useful if the path of propagation is not likely to cause a significant readjustment of the stress field prior to branching.

Experimental observations of the speed of crack propagation at branching suggest that elastodynamic effects are important. For planar crack propagation, these effects were originally investigated by Yoffe [38] who attempted to explain the branching of cracks by analyzing a dynamic steady-state problem in which a crack of fixed length travels at constant velocity under the influence of uniaxial tension. Craggs [39] considered the same problem for a semi-infinite crack. In [38] and [39] it is found that the maximum $\sigma_{\theta\theta}$· stress moves out of the plane of crack propagation and acts at angle of 60° to the direction of crack propagation when the crack velocity exceeds 0.6 times the shear wave velocity. Unfortunately, the crack branching angle and crack tip velocity at branching are consistently lower than the above values (for a survey of the brittle fracture velocities, see Table 1 in [5]). It may be concluded, therefore, that the tendency of the maximum tangential stress $\sigma_{\theta\theta}$ to move out of the plane of crack propagation at higher velocities is at best a contributory factor to the branching event.

The fast growth of cracks in non-uniform stress fields, without any a priori assumption regarding the crack path, has as yet not yielded to current, analytical research efforts. Recent attempts to determine a necessary condition for crack branching have adopted an alternative approach in which states prior to and after branching are compared. The comparison requires expressions for the elastodynamic field quantities near the tips of the branches. The necessary condition can then be established, for instance, on the basis of the balance of rates of energies.

The particular elastodynamic crack branching studies that have been achieved to date are reviewed in Section 2, together with a brief description of the methods of analysis used. Problems which currently require urgent attention as well as some recent research efforts are described in Section 3.

2. Dynamic crack kinking and bifurcation

The sudden branching of an initially stationary crack gives rise to a self-similar solution if the following situation is assumed: each branching crack tip propagates at a constant velocity from the instant that either an incident plane stress wave strikes the semi-infinite crack tip or that the spatially uniform loading is applied to the semi-infinite crack faces. Then, no characteristic length enters the formulation. The ensuing solution is valid for a crack of finite length up to the time required for a signal to travel from the other crack tip.

Analytical solutions to elastodynamic self-similar crack branching problems in homogeneous, isotropic and elastic brittle solids were obtained recently for the case of antiplane strain deformations. Burgers and Dempsey [40] solved the symmetric crack bifurcation in antiplane strain for a specific angle, while Dempsey et al. [41] solved the mode III crack kinking problem for stress wave loading. Dynamic effects in mode III crack branching were explored by Dempsey et al. [42], in which both symmetric and asymmetric crack bifurcation were treated. The solutions in [40,41] verified the numerical method being applied by Burgers [43]; this method has since been used by Burgers [44] and Burgers and Dempsey [45] to provide numerical solutions to the plane strain crack kinking and bifurcation problems, respectively.

The analytical solution method used in the antiplane strain investigations [40–42] is briefly applied here to the plane strain crack kinking problem (Fig. 1). In a stationary system of polar coordinates (r, θ), the displacement potentials $\phi(r, \theta, t)$ and $\psi(r, \theta, t)$ satisfy uncoupled wave equations

$$\nabla^2\phi = c_L^{-2}\ddot{\phi}, \quad \nabla^2\psi = c_T^{-2}\ddot{\psi}, \qquad (2.1a,b)$$

where $c_L = (\lambda + 2\mu)^{1/2}/\rho^{1/2}$ and $c_T = (\mu/\rho)^{1/2}$ are the longitudinal and transverse wave velocities, respectively, and ∇^2 is the two-dimensional Laplacian. The displacement and stress components are expressed in terms of ϕ and ψ in the usual manner [46]. For geometrical configurations and initial-boundary conditions that do not introduce a fixed characteristic length, and for loading that has a step function dependence in time, the quantities $\ddot{\phi}$ and $\ddot{\psi}$ are self-similar; that is, functions of $s = r/t$ and θ only. This crucial observation allows the Chaplygin transformations defined here by

$$\beta_L = \cosh^{-1}(c_L/s), \quad \beta_T = \cosh^{-1}(\beta_T/s), \qquad (2.2a,b)$$

to be used within the cylindrical longitudinal $(s < c_L)$ and transverse $(s < c_T)$ waves, and the governing equations of motion (2.1a,b) are reduced to the following Laplace equations

$$\frac{\partial^2\ddot{\phi}}{\partial\beta_L^2} + \frac{\partial^2\ddot{\phi}}{\partial\theta^2} = 0, \quad \frac{\partial^2\ddot{\psi}}{\partial\beta_T^2} + \frac{\partial^2\ddot{\psi}}{\partial\theta^2} = 0. \qquad (2.3a,b)$$

The real transformations given by (2.2a,b) map the interior of each physical cylindrical region into the interior of semi-infinite strips in the $\gamma_L = \beta_L + i\theta$ and $\gamma_T = \beta_T + i\theta$

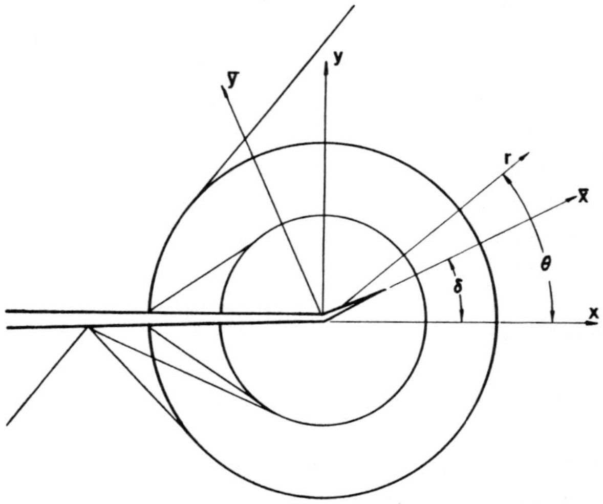

Figure 1. Skew crack geometry and wave front pattern for an incident longitudinal stress wave diffracted by a semi-infinite crack.

-planes. The solutions to Laplace's equations in the strips can be written as the real parts of analytic functions of the complex variables γ_L and γ_T. It is more convenient, however, to conformally map the γ_L and γ_T-planes on the upper halves of complex $\zeta_L = \xi_L + i\eta_L$ and $\zeta_T = \xi_T + i\eta_T$-planes, respectively. The boundaries of the strips are mapped onto the real axes. Appropriate conformal mappings are given by the Schwarz-Christoffel transformation. The mapping for γ_L can be written as [41]

$$\gamma_L = (1-\kappa)\cosh^{-1}\left(\frac{1-\zeta_L(\xi_N)_L}{\gamma_L-(\xi_N)_L}\right) + (1+\kappa)\cosh^{-1}\left(\frac{1+\zeta_L(\xi_M)_L}{\zeta_L+(\xi_M)_L}\right) + i\pi, \tag{2.4}$$

and the corresponding expression for γ_T is obtained by replacing subscripts L by subscripts T in (2.4). In the ζ_L- and ζ_T-planes

$$\ddot{\phi}(\xi_L, \eta_L) = \mathrm{Re}[F_L(\zeta_L)], \quad \ddot{\psi}(\xi_T, \eta_T) = \mathrm{Re}[F_T(\zeta_T)], \tag{2.5a,b}$$

where the analytic functions $F_L(\zeta_L)$ and $F_T(\zeta_T)$ are coupled along the real axes of the half-planes by restrictions which stem from the conditions along the faces of the primary crack and the crack branch, and along the wave fronts (see [47] for more details). This coupling is a major source of difficulty in the plane strain studies. A further complication lies in the fact that the relationships between ζ_L and ζ_T and the physical coordinate r/t and θ given by (2.4) are generally not invertible for problems involving dynamic crack branching. The incorporation of the appropriate singular behavior in a particular problem through the analytic functions in (2.5) hinges upon establishing the relations between small distances from the crack tip in the physical plane, in the γ-planes, and in the ζ-planes. For this reason, asymptotic inversions of the above type play a key role in the kinking and bifurcation studies presented in [41] and [42].

The crucial part of the analysis is the derivation of $F_L(\zeta_L)$ and $F_T(\zeta_T)$. The completeness of these functions must be established: terms which violate the boundary conditions or introduce inadmissible singularities at loading discontinuities or the moving crack tip, for instance, must be excluded. Achenbach and Varatharajulu [48] (kinked crack propagation) and Achenbach [49,50] (crack bifurcation) first attempted the work in [40–42] a decade ago and while they helped to establish the suitability and capability of the method under discussion with regard to self-similar dynamic crack branching problems, they did not specify complete solutions for the analytic functions involved. With recent progress in understanding the mathematical methods involved in [40–42], the level of understanding of the above method has now reached the point where it is realistic to pursue analytical solutions to the associated plane strain problems.

An alternative approach to the above self-similar problems was developed by Burgers [43–45], following the work of Freund [51–53]. The essential details of this approach are presented here in terms of the plane strain kinked crack propagation solution obtained by Burgers [44]; shown in Fig. 1 is the associated wavefront pattern for a plane longitudinal stress wave incident on a stationary semi-infinite crack. The displacements are given by u_x, u_y in the x, y directions, respectively. Plane strain dictates that $u_z \equiv 0$. The relevant stress components needed for solving the boundary value problem are σ_{xx}, σ_{xy} and σ_{yy}. A bar over a subscript indicates that a quantity is represented in the $\bar{x}-\bar{y}$ coordinate system, e.g. $u_{\bar{y}}$ is the displacement in the $\bar{y}$-direction. Consider now that the stationary semi-infinite crack is subjected to loading that generates stresses which are self-similar in r and t. The solution method in [44] relies upon solutions to the fundamental dynamic problems described in the following paragraphs A and B involving propagating singular stress sources in a linearly elastic full space which is considered to be at rest for all $t \leqslant 0$ with zero displacements.

A. Normal and tangential displacement discontinuities which appear at time $t = 0$ with no crack present grow linearly with time and propagate out along the ultimate skew crack path at a polar angle δ and with uniform velocity u. The normal displacement discontinuity is defined as having zero shear stress $\sigma_{\bar{x}\bar{y}}$ and zero normal displacement $u_{\bar{y}}$ on $|\bar{x}| < \infty$, $\bar{y} = 0$ except that for $t > 0$, $u_{\bar{y}} = t$ on $0 \leqslant \bar{x} \leqslant ut$, $\bar{y} = 0$. Similarly, the tangential displacement discontinuity has zero normal stress $\sigma_{\bar{y}\bar{y}}$ and zero tangential displacement $u_{\bar{x}}$ on $|\bar{x}| < \infty$, $\bar{y} = 0$ except for $t > 0$, $u_{\bar{x}} = t$ on $0 \leqslant \bar{x} \leqslant ut$, $\bar{y} = 0$. Details of the solution techniques for the above problems are provided in [44]; the resultant stress components are denoted here by $\sigma^{N}(r/t, \theta; u)$ and $\sigma^{T}(r/t, \theta; u)$, respectively, where $\sigma \equiv (\sigma_{xx}, \sigma_{xy}, \sigma_{yy})$. The velocity u is used as a superposition parameter; that is, a distribution of displacement discontinuities with velocities u ranging from zero to v, the speed of propagation of the skew crack tip. For the problems considered, the resultant stress distributions due to this superposition must be self-similar and this implies that the form of each displacement discontinuity must generate self-similar stresses and must therefore grow linearly with time.

B. Normal and shear point loads grow linearly with time and propagate from the crack tip along the stationary semi-infinite crack faces (with no branching crack present) at constant velocity w. The boundary conditions for the normal point load problem are given by zero shear stress σ_{xy} on $|x| < \infty$, $y = 0$ while the normal stress $\sigma_{yy} = t\delta(wt + x)$ on $y = 0$ if $x < 0$; if $x > 0$ on $y = 0$ the normal displacement u_y is zero. Similarly, the shear point load problem has zero normal stress σ_{yy} on $|x| < \infty$, $y = 0$ while the shear stress $\sigma_{xy} = t\delta(wt + x)$ on $y = 0$ if $x < 0$; if $x > 0$ on $y = 0$ the tangential displacement u_x is zero. Solution details are again provided in [44]; the resultant stress components are denoted here by $\sigma^{NP}(r/t, \theta; w)$ and $\sigma^{sP}(r/t, \theta; w)$, respectively. The velocity w is also used as a superposition parameter, being used to cancel stresses on the semi-infinite or old crack line due to the displacement discontinuities described in paragraph A.

Note that it is the stresses σ_{yy} and σ_{xy} due to $\sigma^{N}(r/t, \theta; u)$ and $\sigma^{T}(r/t, \theta; u)$ which must be cancelled on the negative x-axis ($\theta = \pm\pi$, $r > 0$). These stresses are of the form $f(r/t)$ for $\theta = \text{const}$. This means that a fixed stress level propagates out at a constant speed $w = r/t$ along a line $\theta = \text{const}$. The following argument used by Freund [51] is adopted in [44]: "A stress level $f(w)$ radiates out at the speed w for $t > 0$, the speed w having an arbitrary value between zero and the longitudinal wave speed. The x-coordinates at time t of the stress levels moving with speeds w and $w + \mathrm{d}w$ are $-wt$ and $-(w + \mathrm{d}w)t$, respectively. Thus, to first order terms in the infinitesimal $\mathrm{d}w$, the resultant force due to all stress levels with speeds between w and $w + \mathrm{d}w$ is $tf(w)\mathrm{d}w$ and this force acts at $x = -wt$."

The superposition can now be accomplished by integrating over the range of w. The stresses due to the propagating displacement discontinuities (described in paragraph A), but with a semi-infinite traction-free stationary crack along the negative x-axis are

$$\sigma^{NC(TC)}(r/t, \theta; u) = \sigma^{N(T)}(r/t, \theta; u) - \int_0^{c_L} \{\sigma_{yy}^{N(T)}(w, \pi; u)\sigma^{NP}(r/t, \theta; w)$$

$$+ \sigma_{xy}^{N(T)}(w, \pi; u)\sigma^{SP}(r/t, \theta; w)\}\mathrm{d}w, \tag{2.6}$$

where the superscript nc(tc) indicates a normal (tangential) displacement discontinuity appearing at the origin of a full-space with a traction free semi-infinite crack along the negative x-axis (see Figs. 2(a,b)).

The moving normal and tangential displacement discontinuity solutions obtained in (2.6) and shown in Figs. 2(a,b) are now superimposed over some as yet unknown distributions $F^{N}(u)$ and $F^{T}(u)$, respectively, using their velocity u as the superposition parameter. The stress at any point is finally given by

$$\sigma(r/t, \theta) = \int_0^v \{\sigma^{NC}(r/t, \theta; u)F^N(u) + \sigma^{TC}(r/t, \theta; u)F^T(u)\}du, \tag{2.7}$$

where v is the speed of propagation of the branching crack tip. To solve the problem of interest, the stresses due to the above superposition along the kinked crack line are set equal to the stresses applied along this line. These applied stresses may take two forms. If the only loading is due to applied tractions on the new crack faces, the stresses (2.7) are set equal to the applied stresses. If the loading is applied elsewhere (not on the new crack faces), the problem with the desired loading is first solved without the new crack. The relevant stresses due to (2.7) are then set equal to the negative of the traction components from the latter solution along the new crack faces. By superposition the original problem will then be solved. In either case, two coupled singular integral equations result. The numerical solution of these integral equations is discussed further in [44].

For an important range of kinking angles ($0° < \delta < 45°$), it turns out that the elastodynamic crack kinking stress intensity factors are affected more by the loading of the new crack faces than by the wedge geometry at the initial crack tip. This observation then suggests that a suitable first approximation would be to ignore the wedge geometry altogether, and to compute the elastodynamic stress intensity factors by considering a crack propagating in its own plane provided, however, that the new faces are subjected to traction corresponding to those of the branched crack. Achenbach et al. [54] considered such an approximation and compared the mode III results obtained with the exact results obtained by Dempsey et al. [41]. The range of kinking angles δ for which the approximation gives good results turns out to be surprisingly large: for $0 \leqslant \delta \leqslant \pi/2$, the error increases as δ increases, but is always less than 10%; for $\delta < 40°$, the error is always less than 2% – which seems remarkable. Prompted by the effectiveness of the approximation in mode III, the plane strain crack kinking problem analyzed numerically by Burgers [44] was solved in the same way. While not as effective, the results of [54] showed very satisfactory agreement with the numerical results of [44] for kinking angles δ up to $\delta < 40°$.

The success of the approximation in [54] was somewhat surprising, particularly for mode III deformations. Kuo and Achenbach [55] investigated the mathematical foundation of the approximation used in [54] and showed that the approximate results correspond to the zeroeth order terms in a perturbation procedure for small kinking angles δ,

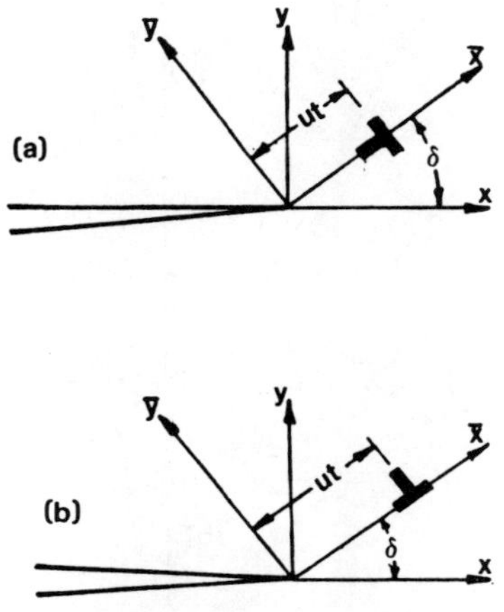

Figure 2. (a) Normal and (b) tangential displacement discontinuities propagating out of a semi-infinite crack tip at a constant velocity u along the kinked crack path.

regardless of whether or not the problems have self-similar field variables. The reason for the particularly good agreement in mode III was also revealed by the perturbation method adopted in [55]. For mode III cases, displacement solutions to the first order system of equations are symmetric with respect to the crack plane, and hence first order contributions to mode III stress intensity factors vanish. Consequently, in the approach of [54], which corresponds to the zeroeth order approximation, the mode III elastodynamic stress intensity factors are accurate to $O(\delta^2)$. For the in-plane problem, the first order mixed mode I–II elastodynamic stress intensity factors do, however, not vanish, and therefore the results of [54] are accurate only to $O(\delta)$.

The computation of the elastodynamic stress intensity factors for arbitrary kinking angles δ (for $\delta < 40°$) and arbitrary (time varying) crack tip speeds are now made possible by the approximation adopted in [54,55], and the work of Freund [51,52] and Kostrov [56], who treated crack propagation with variable velocity under general loading conditions. Achenbach and Kuo [57] investigated conditions for kinking of an initially stationary crack under stress wave loading with the body in a state of static pre-stress. As a criterion for crack kinking the authors chose the maximum energy release rate criterion [58] and used the work in [54–56]. It is shown in [57] that under step-stress wave loading the crack tip speed is zero at the onset of fracture ($t = t_f$) – the fields near the crack tip need a time t_f to build up to a critical state before kinking may take place under an angle δ. As a consequence of this result it is concluded in [57] that the conditions for the onset of crack kinking and for the computation of t_f and δ are completely defined by the elastodynamic field around the original crack tip; for an incident antiplane wave the crack will generally not kink; for an incident in-plane wave the crack may kink depending on the state of pre-stress and the magnitude and angle of incidence of the incident wave. Both t_f and δ can then be determined.

From the comparison of stress intensity factors in Figs. 3(a,b) it is obvious that for plane strain crack bifurcation there is an interaction between the two crack tips. Under a

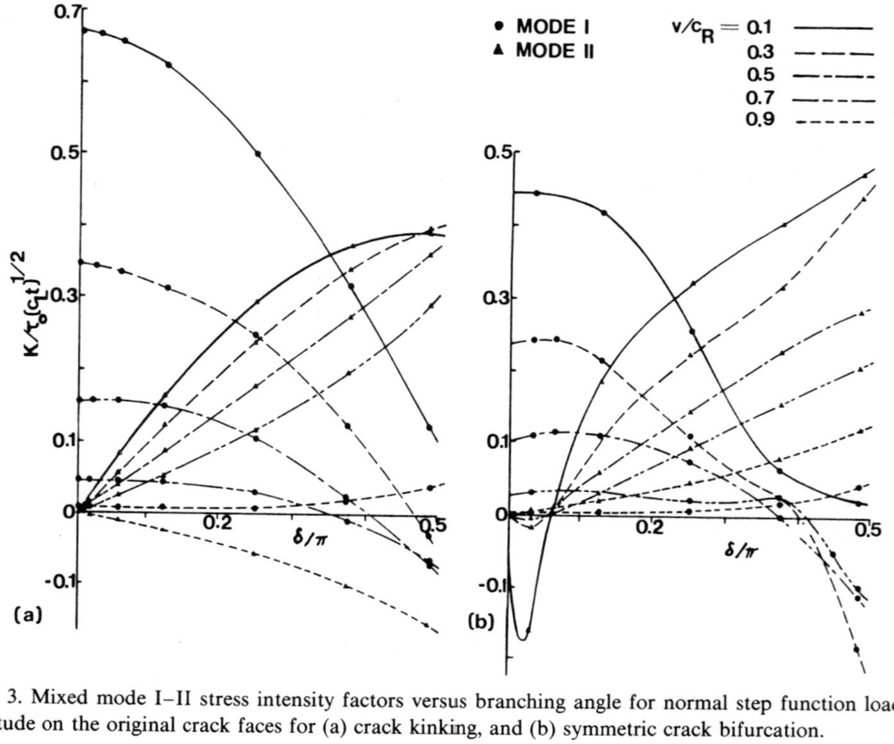

Figure 3. Mixed mode I–II stress intensity factors versus branching angle for normal step function loading of magnitude on the original crack faces for (a) crack kinking, and (b) symmetric crack bifurcation.

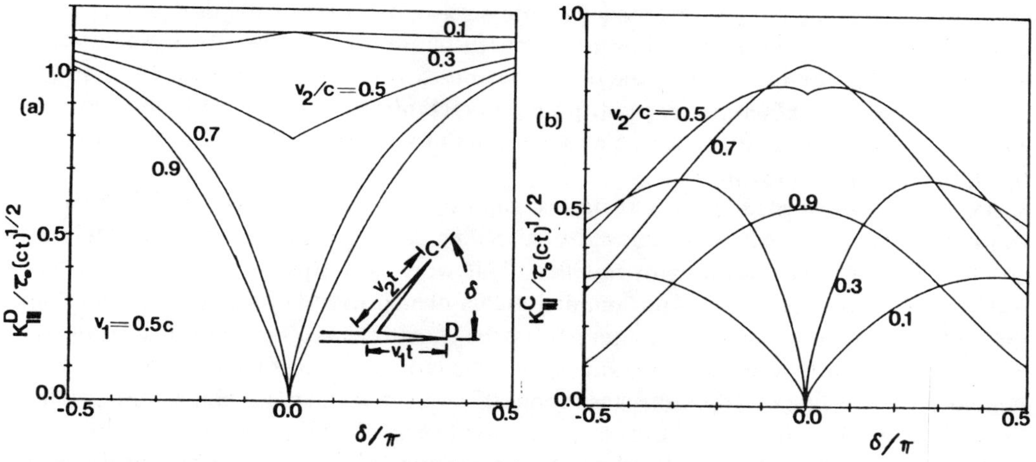

Figure 4. Mode III stress intensity factors versus branching angle for step-stress wave loading (normal incidence) of magnitude: (a) the unbranched semi-infinite crack tip D propagating at a velocity $v_1 = 0.5c$; (b) the secondary branch (crack tip C) propagating at a velocity v_2 under an angle with the primary crack plane.

range of loading conditions a maximum in K_{I} occurs for bifurcation angles well within the experimentally observed range (see Section 1) while for the same angles K_{II} is approximately zero. No trends as regards conditions for branching are yet clear, but this is more than likely a product of the less than realistic models that have been solved to date.

Recent elastodynamic mode III results by Dempsey et al. [42] indicate that the non-symmetrical bifurcation of cracks may have very interesting effects (keep in mind, however, that mode III solutions cannot be used to interpret or assess the corresponding in-plane physical behavior – see the experimental study of crack propagation in anti-plane shear by Knauss [59]). It is found in [42] that a branch cannot propagate almost parallel to the primary crack unless it has a velocity almost equal to that of the primary crack. The crack tip stress intensity factor of a secondary branch propagating under an angle δ with the original crack is relatively close to but less than that of the primary crack – as illustrated in Figs. 4(a,b). In other words, continuous branching with very little crack speed alteration is possibly a necessary condition for successful crack branching; this statement is supported by experimental evidence in [1,5,6,17,22].

3. Future research

The problems that must be solved to obtain a reasonable understanding of the branching phenomena are, no doubt, very difficult. Presently, the only ones that appear to be amenable to treatment are those that are reduced to an idealized plane strain situation. To arrive at conditions for crack branching, however, less idealized problems involving non-simultaneous non-contiguous branching, branching at different locations on the (curved) crack front, branching of a propagating crack into a stress field that is not self-similar, tunnelling, etc., should be solved.

The crack branching mechanism demonstrated by Ravi-Chandar and Knauss [4,5] involves the coalescence and interaction of non-contiguous microcracks with the primary crack. Macroscopically, the propagating crack is observed to branch with each branch deviating smoothly from the original crack path. Models of either phenomena, even for the case plane strain, are very difficult to solve. However, it is up to the analyst to provide fundamental solutions to problems involving dynamically interacting multiple crack geometries and the branching of non-stationary cracks into stress fields that are not self-similar. In this context, the following problem is illustrative of the direction in which future research must focus. Consider a stationary semi-infinite crack that is subjected to dynamic loading conditions. At some point ahead of the stationary crack tip, at some time after the stress waves have passed that point, a crack initiates from it with one end fixed while the other propagates at an angle to, or smoothly deviates along a curved path away from, the primary crack plane.

The development of numerical methods using the finite element and finite difference techniques that are capable of treating the dynamic crack growth process accurately and efficiently is an important area of research. This work is needed so that the directional stability of propagating cracks in bounded media, which have boundaries and interfaces that reflect the stress waves emanating from the moving crack tip, can be investigated. There are currently very few investigations concerning these topics in the literature. Kinked crack propagation under the application of transient antiplane and in-plane incident waves has, however, been investigated by Aboudi [60] using a finite difference technique. Unfortunately, the benchmark solutions in [41,43,44] were not available for comparison at the time.

Acknowledgements

The second author (P.B.) would like to gratefully acknowledge the support of the National Science Foundation, Solid Mechanics Program under the grant MEA-8306644.

References

[1] H. Schardin, in *Fracture*, edited by B.I. Averbach, D.K. Felbeck, G.T. Hahn and D.A. Thomas, The M.I.T. Press, Cambridge (1959) 297–330.
[2] A-B.J. Clark and G.R. Irwin, *Experimental Mechanics* 6 (1966) 321–330.
[3] K. Ravi-Chandar and W.G. Knauss, *International Journal of Fracture* 25 (1984) 247–262.
[4] K. Ravi-Chandar and W.G. Knauss, *International Journal of Fracture* 26 (1984) 65–80.
[5] K. Ravi-Chandar and W.G. Knauss, *International Journal of Fracture* 26 (1984) 141–154.
[6] K. Ravi-Chandar and W.G. Knauss, *International Journal of Fracture* 26 (1984) 189–200.
[7] J.E. Field, *Contemporary Physics* 12 (1971) 1–31.
[8] R.D. Streit and I. Finnie, *Experimental Mechanics* 20 (1980) 17–23.
[9] B. Cotterell and J.R. Rice, *International Journal of Fracture* 16 (1980) 155–169.
[10] G.M. Boyd, *Brittle Fracture in Steel Structures*, Butterworth and Co. (1970).
[11] J.W. Dally and W.L. Fourney, *Proceedings of the Eighteenth Rock Mechanics Symposium*, Keystone, CO (1977).
[12] H.P. Rossmanith and W.L. Fourney, *Rock Mechanics* 14 (1982) 209–233.
[13] G.T. Hahn, R.G. Hoagland and A.R. Rosenfield, in *Fracture*, University of Waterloo Press, Vol. 2 (1977) 1333–1338.
[14] Z.T. Bieniawski, *International Journal of Fracture* 4 (1968) 415–430.
[15] J.W. Dally, *Experimental Mechanics* 19 (1979) 349–361.
[16] J. Congleton, in *Dynamic Crack Propagation*, edited by G.C. Sih, Noordhoff, Leyden (1973) 427–438.
[17] A.S. Kobayashi, B.G. Wade, W.B. Bradley, and S.T. Chiu, *Engineering Fracture Mechanics* 6 (1974) 81–92.
[18] M. Ramulu, A.S. Kobayashi and B.S.-J. Kang, in *Fifteenth National Symposium on Fracture Mechanics* (1982).
[19] M. Ramulu and A.S. Kobayashi, *Experimental Mechanics* 23 (1983) 1–9.
[20] A.S. Kobayashi, in *Proceedings, Workshop on Dynamic Fracture*, California Institute of Technology, Pasadena (1983).
[21] J.F. Kalthoff, in *Dynamic Crack Propagation*, edited by G.C. Sih, Noordhoff, Leyden (1973) 449–458.
[22] S. Aoki and M. Sakata, in *Symposium of Mechanical Behavior of Materials*, Society of Materials Science (Japan), Kyoto (1974) 11–8(1).
[23] F. Erdogan and G.C. Sih, *Journal of Basic Engineering* 85 (1963) 519–527.
[24] G.C. Sih, *Journal of Applied Mechanics* 32 (1965) 51–58.
[25] E. Smith, *Journal of the Mechanics and Physics of Solids* 16 (1968) 329–336.
[26] M.A. Hussain, S.L. Pu and J. Underwood, ASTM-STP-560 (1974) 2–28.

[27] K. Palaniswamy and W.G. Knauss, On the Problem of Crack Extension in Brittle Solids under General Loading, Report SM 74-8, Graduate Aeronautical Laboratories, California Institute of Technology, Pasadena (1974).
[28] C.-H. Wu, *Journal of Applied Mechanics* 45 (1978) 553–558.
[29] K.K. Lo, *Journal of Applied Mechanics* 45 (1978) 797–802.
[30] K. Hayashi and S. Nemat-Nasser, *Journal of Applied Mechanics* 48 (1981) 520–524.
[31] W.K. Wilson and J. Cherepko, *International Journal of Fracture* 22 (1983) 303–315.
[32] N.V. Banichuk, *Akademiya Nauk SSSR Izvestiya. Mekhanika Tverdogo Tela* (Mechanics of Solids) 7 (1970) 130–137 (in Russian).
[33] B.L. Karihaloo, L.M. Keer, S. Nemat-Nasser and A. Oranratnachai, *Journal of Applied Mechanics* 48 (1981) 515–519.
[34] S. Melin, *International Journal of Fracture* 23 (1983) 37–45.
[35] E. Viola and A. Piva, *Engineering Fracture Mechanics* 19 (1984) 1069–1084.
[36] S.K. Maiti and R.A. Smith, *Journal of the Mechanics and Physics of Solids* 31 (1983) 389–403.
[37] S.K. Maiti and R.A. Smith, *Engineering Fracture Mechanics* 19 (1984) 793–804.
[38] E.H. Yoffe, *Philosophical Magazine* 42 (1951) 739–750.
[39] J.W. Craggs, *Journal of the Mechanics and Physics of Solids* 8 (1960) 66–75.
[40] P. Burgers and J.P. Dempsey, *Journal of Applied Mechanics* 49 (1982) 366–370.
[41] J.P. Dempsey, M.-K. Kuo and J.D. Achenbach, *Wave Motion* 4 (1982) 181–190.
[42] J.P. Dempsey, M.-K. Kuo and D.L. Bentley, Dynamic Effects in Mode III Crack Bifurcation, *International Journal of Solids and Structures* (submitted for publication).
[43] P. Burgers, *Journal of Applied Mechanics* 49 (1982) 371–376.
[44] P. Burgers, *International Journal of Solids and Structures* 19 (1983) 735–752.
[45] P. Burgers and J.P. Dempsey, *International Journal of Solids and Structures* 20 (1984) 609–618.
[46] J.D. Achenbach, *Wave Propagation in Elastic Solids*, North-Holland/American Elsevier, Amsterdam/New York (1973).
[47] J.D. Achenbach and R.P. Khetan, *Journal of Elasticity* 9 (1979) 113–129.
[48] J.D. Achenbach and V.K. Varatharajulu, *Quarterly of Applied Mathematics* 32 (1974) 123–135.
[49] J.D. Achenbach, in *Prospects of Fracture Mechanics*, edited by G.C. Sih, H.C. van Elst and D. Broek, Noordhoff, Leyden (1974) 319–336.
[50] J.D. Achenbach, *International Journal of Solids and Structures* 11 (1975) 1301–1314.
[51] L.B. Freund, *Journal of the Mechanics and Physics of Solids* 20 (1972) 141–152.
[52] L.B. Freund, *Journal of the Mechanics and Physics of Solids* 21 (1973) 47–61.
[53] L.B. Freund in *Mechanics Today*, edited by S. Nemat-Nasser, 3 (1976) 55–91.
[54] J.D. Achenbach, M.-K. Kuo and J.P. Dempsey, *International Journal of Solids and Structures* 20 (1984) 395–410.
[55] M.-K. Kuo and J.D. Achenbach, Perturbation Method to Analyze the Elastodynamic Field Near a Kinked Crack, *International Journal of Solids and Structures* (to be published).
[56] B.V. Kostrov, *International Journal of Fracture* 11 (1975) 47–56.
[57] J.D. Achenbach and M.-K. Kuo, Conditions for Crack Kinking Under Stress Wave Loading, *Engineering Fracture Mechanics* (to be published).
[58] L.B. Freund, *Journal of Elasticity* 2 (1972) 341–349.
[59] W.G. Knauss, *International Journal of Fracture Mechanics* 6 (1970) 183–187.
[60] J. Aboudi, *Computer Methods in Applied Mechanics and Engineering* 15 (1978) 181–199.

Résumé

On présente un état de la question relatif aux recherches analytiques et numériques récemment effectuées ou en cours sur le phénomène d'arborescence de fissures dynamiques dans des solides fragiles. On passe d'abord en revue les constatations expérimentales de base concernant la physique du mécanisme de l'arborescence. Ensuite, est résumé un travail élastostatique relatif à l'arborescence d'une fissure sous un angle arbitraire et les critères de rupture d'une fissure en arborescence. On met l'accent sur les idéalisations auxquelles conduisent les modèles mathématiques étudiés. On discute les schémas analytiques et numériques qui ont récemment conduit à des solutions au phénomène de branchment d'une fissure dynamique dans des conditions comparables. On discute également une approximation de type perturbation applicable aux conditions initiales aux limites et par laquelle la contrainte artificielle d'auto-similarité peut être éliminée et le phénomène d'arborescence d'une fissure sous vélocité variable peut être traité. Finalement, on identifie de nouvelles directions et les besoins pour des recherches futures.

International Journal of Fracture 27 (1985) 215–227.

Dynamic fields generated by rapid crack growth

J.D. ACHENBACH, Z.L. LI and N. NISHIMURA

Department of Civil Engineering, Northwestern University, Evanston, IL, 60201, USA

(Received September 1, 1984)

Abstract

Stresses and strains near a rapidly propagating crack tip are affected by the mass density of the material. This paper starts with a brief summary of analytical results for near-tip dynamic fields as predicted by linear elastic fracture mechanics. Next, exact expressions are derived for dynamic crack-line strains, for mode-III crack propagation in a nonlinear elastic material and in an elastic perfectly-plastic material. These expressions are valid on the crack line from the moving crack tip to the moving boundary with the region of linearly elastic deformation. For steady-state crack growth, a critical strain criterion is used to compute the relation between external load and crack tip speed. The required external load increases with crack-tip speed.

1. Introduction

Dynamic fracture mechanics applies when inertia forces must be included in the governing equations for an accurate analysis of crack-growth initiation, crack propagation and crack arrest. Inertia forces are important when cracked bodies are subjected to high-rate external loads, such as impact loads and explosive charges, or when a crack propagates very fast. Spalling is an example of a dynamic fracture phenomenon caused by rapid load application. In rapid crack propagation inertia forces are important because they affect the fields of stress and deformation near the propagating crack tip, and hence the conditions for propagation and arrest. This paper is concerned with dynamic fracture mechanics for rapid crack growth.

The paper starts with a brief review of available analytical results for the dynamic near-tip fields according to linear elastic fracture mechanics. In Section 4 a nonlinear elastic model is considered, and some new exact results are presented for dynamic mode-III fields on the crack line. A rather interesting feature is that the strains as well as the stresses remain bounded. In the last section of the paper, crack propagation in an elastic, perfectly-plastic material is considered, in order to investigate the combined effect of plastic deformation and mass density.

2. Dynamic linear elastic fracture mechanics

For essentially brittle fracture, near-tip dynamic effects have been investigated extensively on the basis of linear elastic fracture mechanics. By now, several papers have reviewed the computation of elastodynamic stress intensity factors, and they have discussed dynamic effects on the fracture criterion of the balance of rates of energies, [1–3]. Analytical methods have yielded three general results for the dynamic fields near a rapidly propagating crack tip: (1) the asymptotic form of the near-tip fields; (2) an expression for the elastodynamic stress intensity factor in terms of the corresponding quasi-static stress-intensity factor, but only for the special case of a propagating semi-infinite crack; and (3) a general expression for the flux of energy into a propagating crack tip. In addition, many numerical results have been obtained, both by the finite element and the finite difference method.

The geometry that will be considered here is shown in Fig. 1. The x_3-axis of a stationary coordinate system is parallel to the crack front, and x_1 points in the direction of crack growth. The position of the crack tip is defined by $x_1 = a(t)$. A moving coordinate system (x, y, z) is centered at the crack tip, with its axes parallel to the x_1, x_2 axes. We also define a moving system of polar coordinates (r, θ), centered at the moving crack tip.

Let us first examine the near-tip behavior of the stress components τ_θ (mode I) and $\tau_{\theta z}$ (mode III), in the polar coordinate systems shown in Fig. 1, for both the elastostatic and the elastodynamic cases, and for a stationary as well as a running crack tip. For the elastostatic case we have

$$\tau_\theta = (2\pi)^{-1/2} r^{-1/2} K_{\mathrm{I}}(t) T_\theta^{\mathrm{I}}(\theta),$$

$$\tau_{\theta z} = (2\pi)^{-1/2} r^{-1/2} K_{\mathrm{III}}(t) T_{\theta z}^{\mathrm{III}}(\theta). \tag{2.1a,b}$$

In these expressions $T_\theta^{\mathrm{I}}(0) = T_{\theta z}^{\mathrm{III}}(0) \equiv 1$. For the elastostatic case the time t is a parameter, which may enter via the loading, or via the length of the crack if the crack is propagating. For a stationary crack tip, the elastodynamic expressions analogous to (2.1a) and (2.1b) are completely equivalent in form. The same angular distributions $T_\theta^{\mathrm{I}}(\theta)$ and $T_{\theta z}^{\mathrm{III}}(\theta)$ apply. We use the notation

$$k_{\mathrm{I}}(t) \text{ and } k_{\mathrm{III}}(t) \tag{2.2a,b}$$

for the elastodynamic stress intensity factors.

The nature of the angular distributions changes for the elastodynamic case of a propagating crack tip. Now the speed of crack propagation, $\dot{a}(t)$, enters the expressions for the angular distributions. We use the notations $T_\theta^{\mathrm{I}}(\theta, \dot{a})$ and $T_{\theta z}^{\mathrm{III}}(\theta, \dot{a})$, where $T_\theta^{\mathrm{I}}(0, \dot{a}) = T_{\theta z}^{\mathrm{III}}(0, \dot{a}) \equiv 1$. General expressions for $T_\theta^{\mathrm{I}}(\theta, \dot{a})$ and $T_{\theta z}^{\mathrm{III}}(\theta, \dot{a})$ have been derived by Achenbach and Bazant [4]. For the stress intensity factors we use the notations $k_{\mathrm{I}}(t, \dot{a})$ and $k_{\mathrm{III}}(t, \dot{a})$. The function $T_\theta^{\mathrm{I}}(\theta, \dot{a})$ is independent of the overall geometry and the loading. It is of note that the maximum value of $T_\theta^{\mathrm{I}}(\theta, \dot{a})$ moves out of the plane $\theta = 0$ (the plane of crack propagation) as $\dot{a}(t)$ increases beyond a certain value. Generally $k_{\mathrm{I}}(t, \dot{a})$ is much more difficult to compute than the corresponding quasi-static stress intensity factor.

For a semi-infinite crack, an interesting relation between $k_{\mathrm{I}}(t, \dot{a})$ and the corresponding quasi-static stress-intensity factor $K_{\mathrm{I}}(t)$ was derived by Freund, see e.g. [2]. He showed

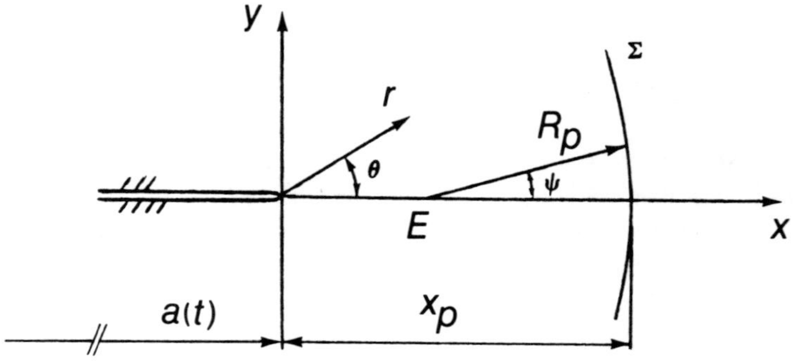

Figure 1. Propagating crack tip.

that

$$k_{\mathrm{I}}(t, \dot{a}) = f(\dot{a}) K_{\mathrm{I}}(t), \tag{2.3}$$

where $f(\dot{a})$ depends on the crack-tip speed and the characteristic wave speeds of the material. The function $f(\dot{a})$ is complicated. For the mode-III case the corresponding relation is, however, of the simple form

$$k_{\mathrm{III}}(t, \dot{a}) = (1 - M)^{1/2} K_{\mathrm{III}}(t), \tag{2.4}$$

where M is the Mach number

$$M = \dot{a}/c_{\mathrm{T}} = \dot{a}/(\mu/\rho)^{1/2}. \tag{2.5}$$

Unfortunately, simple relations of the kinds (2.4) and (2.5) hold only for semi-infinite cracks. For cracks of finite lengths numerical methods are the only practical way to compute elastodynamic stress-intensity factors.

The conditions governing crack tip motion can be expressed in terms of $k_{\mathrm{I}}(t, \dot{a})$ and an experimentally determined critical value that is assumed to be a property of the material. In conventional (quasi-static) LEFM one has $K_{\mathrm{I}} = K_{\mathrm{c}}$ as the condition for crack instability. In the dynamic generalization of LEFM, K has two counterparts. First, for the initiation of crack growth

$$k_{\mathrm{I}}(t, 0) = K_{\mathrm{d}}(\dot{\sigma}), \tag{2.6}$$

where $\dot{\sigma}$ represents the loading rate. For perfectly brittle fracture $K_{\mathrm{c}} = K_{\mathrm{d}}$. Similarly, for a propagating crack

$$k_{\mathrm{I}}(t, \dot{a}) = K_D(\dot{a}), \tag{2.7}$$

where K_D is known as the dynamic fracture toughness.

A more general approach is based on the observation that a propagating crack tip acts as an energy sink. In terms of the energy release rate $\mathscr{G}$, a necessary condition for crack tip motion can be expressed as $\mathscr{G} = \mathscr{R}(\dot{a})$. Here $\mathscr{R}(\dot{a})$, which is the critical value of $\mathscr{G}$, defines the material's resistance to fracture. It represents the energy dissipated in the fracture processes accompanying crack propagation.

The energy release rate, $\mathscr{G}$, and the flux of energy into the crack tip, $\mathscr{F}$, are related by $\mathscr{F} = \mathscr{G}\dot{a}$. For elastodynamic problems where time enters as an independent parameter, it is quite simple to compute the flux of energy into the crack tip. Detailed discussions can be found in [1] and [2]. In a two-dimensional geometry, $\mathscr{F}$ can be expressed in the form

$$\mathscr{F} = -\frac{\dot{a}^3}{2\mu c_{\mathrm{T}}^2 D(\alpha, \beta)}\left\{(1-\alpha^2)^{1/2}[k_{\mathrm{I}}(t, \dot{a})]^2 + (1-\beta^2)^{1/2}[k_{\mathrm{II}}(t, \dot{a})]^2\right\} + \frac{\dot{a}}{2\mu(1-\beta^2)^{1/2}}[k_{\mathrm{III}}(t, \dot{a})]^2, \tag{2.8}$$

where $D(\alpha, \beta)$ is defined as

$$D(\alpha, \beta) = (\beta^2 - 2)^2 - 4(1-\alpha^2)^{1/2}(1-\beta^2)^{1/2}. \tag{2.9}$$

In (2.8), $\alpha = \dot{a}/c_{\mathrm{L}}$, $\beta = \dot{a}/c_{\mathrm{T}}$, where c_{L} and c_{T} are the speeds of longitudinal and transverse waves, respectively. In terms of the flux of energy into the crack tip the necessary condition for fracture is

$$\mathscr{F} = \dot{a}\mathscr{R}(\dot{a}). \tag{2.10}$$

In purely brittle fracture $\frac{1}{2}\mathscr{R}(\dot{a})$ becomes the specific surface energy.

For the case of mode I fracture (2.8) becomes

$$\mathscr{F} = -\frac{\dot{a}^3(1-\alpha^2)^{1/2}[k_{\mathrm{I}}(t, \dot{a})]^2}{2\mu c_{\mathrm{T}}^2 D(\alpha, \beta)}, \tag{2.11}$$

while slow propagation of the crack tip (i.e. the limit $\dot{a} \to 0$) reduces (2.11) to

$$\mathscr{F} = \frac{c_{\mathrm{L}}^2 [K_{\mathrm{I}}(t)]^2 \dot{a}}{4\mu (c_{\mathrm{L}}^2 - c_{\mathrm{T}}^2)} = \frac{1-\nu^2}{E} [K_{\mathrm{I}}(t)]^2 \dot{a}. \tag{2.12}$$

By combining this result with (2.10), we see that K_{I} equals the toughness for fracture initiation, K_{c}, i.e.,

$$K_{\mathrm{I}} = K_{\mathrm{c}} = \left[\frac{\mathscr{R}(0) E}{1 - \nu^2} \right]^{1/2}. \tag{2.13}$$

Thus, for a slowly propagating crack the usual fracture toughness enters as a material property, and the stress intensity factor equals the fracture toughness K_{c} while the crack propagates. The four relations (2.6), (2.7), (2.11) and (2.12) can be used to establish the following theoretical relation between K_{D} and K_{c}

$$K_{\mathrm{D}} = \left[-\frac{1}{2} \frac{c_{\mathrm{T}}^2}{c_{\mathrm{L}}^2 - c_{\mathrm{T}}^2} \frac{D(\alpha, \beta)}{\alpha^2 (1-\alpha^2)^{1/2}} \frac{\mathscr{R}(\dot{a})}{\mathscr{R}(0)} \right]^{1/2} K_{\mathrm{c}}, \tag{2.14}$$

where, as in (2.13), $\mathscr{R}(0)$ corresponds to the quasi-static critical energy release rate.

3. Governing equations for nonlinear models

For two nonlinear constitutive models some interesting exact analytical results can be obtained on the crack line, but so far only for the mode-III case. In this section we state the governing equations that will be considered.

In the moving coordinate system the equation of motion is

$$\frac{\partial \tau_{xz}}{\partial x} + \frac{\partial \tau_{yz}}{\partial y} = \rho \ddot{w}, \tag{3.1}$$

where ρ is the mass density, $w(x, y, t)$ is the anti-plane displacement, and the material time derivative is defined as

$$(\dot{\ }) = (\partial / \partial t) - \dot{a} (\partial / \partial x). \tag{3.2}$$

Here $\dot{a} = \mathrm{d}a / \mathrm{d}t$ is the speed of the crack tip. The relevant strain components are

$$\gamma_x = \frac{\partial w}{\partial x}, \quad \gamma_y = \frac{\partial w}{\partial y}. \tag{3.3a,b}$$

Two kinds of constitutive relations will be considered: a nonlinear elastic model and an elastic perfectly-plastic model. The two models have in common that the following constraint condition applies

$$\mathscr{T}^2 = \tau_{yz}^2 + \tau_{yz}^2 \leqslant k^2. \tag{3.4}$$

Nonlinear elastic model

For this model the stress components are completely defined by the instantaneous values of the strains. We consider the following special case

$$\tau_{\alpha z} = \mu \frac{\partial w}{\partial x_\alpha}, \quad \mathscr{T} < k; \qquad \tau_{\alpha z} = \frac{k}{\epsilon} \frac{\partial w}{\partial x_\alpha}, \quad \mathscr{T} = k, \tag{3.5a,b}$$

where μ is the shear modulus, $\alpha = x,\ y$ and

$$\epsilon^2 = \left(\frac{\partial w}{\partial x} \right)^2 + \left(\frac{\partial w}{\partial y} \right)^2. \tag{3.6}$$

It is well known that this model cannot be used to obtain the strains for given stresses.

Elastic perfectly-plastic model

For this model k in (3.4) is interpreted as the yield stress in pure shear, and

$$\mathscr{T}^2 = \tau_{xz}^2 + \tau_{yz}^2 = k^2 \tag{3.7}$$

becomes the Huber-Mises yield condition. In the zone of plastic deformation, the strain rates are related to the stresses and the stress rates by

$$\dot{\gamma}_x = \frac{\dot{\tau}_{xz}}{\mu} + 2\dot{\Lambda}\tau_{xz}, \quad \dot{\gamma}_y = \frac{\dot{\tau}_{yz}}{\mu} + 2\dot{\Lambda}\tau_{yz}. \tag{3.8a,b}$$

In (3.8a,b) μ is the shear modulus, and $\dot{\Lambda}$ is a positive function of time and the spatial coordinates, and the dot derivatives are defined by (3.2).

4. Crack-line solutions for the nonlinear elastic model

In this section attention is restricted to constant crack-tip speed and steady-state fields. The material time derivatives then reduce to

$$(\dot{\ }) = -\dot{a}\frac{\partial}{\partial x}, \quad (\ddot{\ }) = \dot{a}^2\frac{\partial^2}{\partial x^2}. \tag{4.1}$$

Relative to the moving coordinate system, the equation of motion (3.1) becomes

$$\frac{\partial \tau_{xz}}{\partial x} + \frac{\partial \tau_{yz}}{\partial y} = \rho\dot{a}^2\frac{\partial^2 w}{\partial x^2}. \tag{4.2}$$

Substitution of the stress-strain relation (3.5b) into the equation of motion (3.1) yields in the zone of nonlinear deformation

$$\left(\frac{\partial w}{\partial y}\right)^2\frac{\partial^2 w}{\partial x^2} - 2\frac{\partial w}{\partial x}\frac{\partial w}{\partial y}\frac{\partial^2 w}{\partial x \partial y} + \left(\frac{\partial w}{\partial x}\right)^2\frac{\partial^2 w}{\partial y^2} = \lambda\left[\left(\frac{\partial w}{\partial x}\right)^2 + \left(\frac{\partial w}{\partial y}\right)^2\right]^{3/2}\frac{\partial^2 w}{\partial x^2}, \tag{4.3}$$

where

$$\lambda = \rho\dot{a}^2/k. \tag{4.4}$$

Equation (4.3) applies inside a zone of nonlinear elasticity at the propagating crack tip. The extent of the zone of nonlinear elasticity along the crack line is defined by $0 \leqslant x \leqslant x_p$. Outside this zone, (3.5a) applies, and the deformation is linearly elastic. The boundary between the zones of nonlinear and linear elasticity is called the transition boundary. It is indicated by Σ in Fig. 1.

A solution to (4.3) valid in the whole zone of nonlinear deformation has not been obtained. It is, however, not difficult to obtain a solution on the crack line in the region $0 \leqslant x \leqslant x_p$. This can be done by an expansion method which has previously been utilized in [5–7]. Taking into account the anti-symmetry of the displacement, we seek a solution of (4.3) for small values of y/x in the following form:

$$w(x, y) = w_1(x)y + w_3(x)y^3 + O(y^5). \tag{4.5}$$

This expression does not preclude the possibility of singular strains, since there are no a-priori restrictions on $w_1(x)$. For future reference we state the following corresponding expansions for τ_{yz} and τ_{xz}

$$\tau_{yz} = s_0(x) + s_2(x)y^2 + O(y^4) \tag{4.6}$$

$$\tau_{xz} = \tau_1(x)y + O(y^3). \tag{4.7}$$

Substitution of (4.5) into (4.3), followed by collection of terms in y of power unity, gives an ordinary differential equation for w_1:

$$w_1 \frac{d^2 w_1}{dx^2} - 2\left(\frac{dw_1}{dx}\right)^2 = \lambda w_1^2 \frac{d^2 w_1}{dx^2}. \tag{4.8}$$

It is convenient to define

$$\gamma_1 = dw_1/dx, \tag{4.9}$$

and to rewrite (4.8) in the form

$$\frac{dw_1}{w_1(1-\lambda w_1)} = \frac{d\gamma_1}{2\gamma_1}. \tag{4.10}$$

Equation (4.10) can be integrated to yield

$$\gamma_1 = \frac{dw_1}{dx} = -\frac{V_0 w_1^2}{(w_1 - 1/\lambda)^2}, \tag{4.11}$$

where the constant V_0 is positive. Integration of (4.11) yields

$$x = -\frac{1}{V_0}\left(w_1 - \frac{2}{\lambda}\ln w_1 - \frac{1}{\lambda^2}\frac{1}{w_1}\right) + C \tag{4.12}$$

where C is another integration constant.

To determine C and V_0 we need conditions at $x = 0$ and $x = x_p$. As shown by Achenbach and Nishimura [7], the following conditions hold at the transition boundary:

$$[[\tau_{\alpha z}]] = 0, \quad [[\nabla w]] = 0, \tag{4.13a,b}$$

where the brackets denote the usual notation for a discontinuity. In conjunction with the linearly elastic stress-strain relations and the expansions (4.5)–(4.7), these conditions

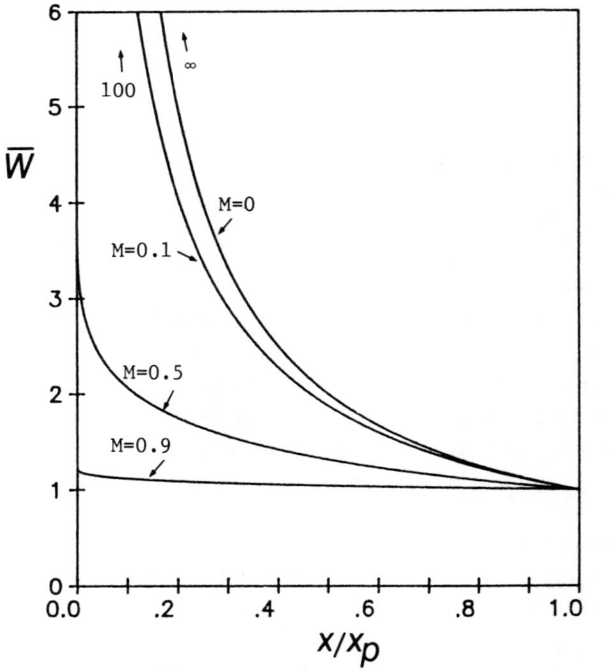

Figure 2. Normalized crack-line strain, $\overline{W} = (\mu/k)\gamma_y|_{y=0}$, versus x/x_p in zone of nonlinear elastic deformation, for various values of M.

imply at

$$y=0,\ x=x_p:\qquad \bar{\tau}_1=\mu\bar{\gamma}_1,\ \text{and}\ \bar{w}_1=k/\mu, \tag{4.14a,b}$$

where

$$(\bar{\tau}_1,\bar{\gamma}_1,\bar{w}_1)=(\tau_1,\gamma_1,w_1)|_{x=x_p}. \tag{4.15a,b,c}$$

Here we have also used that $s_0=k$, which follows by substituting (4.6) and (4.7) into (3.4, with equality sign).

A condition at $x=0$ can be written by considering the singular behavior of the solution. Physical considerations suggest that our solution should be singular in some sense at the crack tip. Equation (4.12) shows, however, that w_1 cannot be unbounded for any finite x. However, we can make γ_1 singular at $x=0$ by setting

$$w_1(0)=1/\lambda, \tag{4.16}$$

where we have used (4.11). With the boundary conditions (4.14) and (4.16) we can determine V_0 and C, to obtain

$$\frac{x}{x_p}=\frac{1-M^4\overline{W}^2+2M^2\overline{W}\ln(M^2\overline{W})}{(1-M^4+2M^2\ln M^2)\overline{W}}, \tag{4.17}$$

where M is the Mach number, defined by (2.5), i.e., $M^2=\dot{a}^2/(\mu/\rho)$, and

$$\overline{W}=w_1/(k/\mu). \tag{4.18}$$

Since the crack tip is in the zone of nonlinear deformation, we conclude by using (3.3b), (4.5) and (4.16) that $1/\lambda>k/\mu$, or by using (4.4): $M^2<1$. Plots of $\overline{W}$ vs x/x_p are given in Fig. 2.

It is noteworthy that $\overline{W}$, and therefore the strain γ_y, remain bounded along the crack line. Also, by letting $M\downarrow 0$ in (4.17), we recover the quasi-static solution

$$\overline{W}=x_p/x. \tag{4.19}$$

Equation (4.17), which is exact, is the main result of this section. It would, however, be desirable to have an explicit expression for x_p in terms of the crack speed and the far-field loading. This can be achieved by assuming that near the crack line the displacement in the zone of linear deformation can be represented by

$$w=\frac{2k_{\mathrm{III}}(\dot{a})}{\mu(1-M^2)^{1/2}}\left(\frac{R}{2\pi}\right)^{1/2}(1-M^2\sin^2\psi)^{1/4}\sin\tfrac{1}{2}\left\{\tan^{-1}\left[(1-M^2)^{1/2}\tan\psi\right]\right\}, \tag{4.20}$$

where R and ψ are polar coordinates centered at the point E, see Fig. 1, and $k_{\mathrm{III}}(\dot{a})$ is the steady-state dynamic stress-intensity factor. The displacements (4.5) and (4.20) can then be connected by using the continuity conditions (4.13a, b) near $x=x_p$.

Equation (4.20) is the well-known steady-state solution for the linearly elastic case. This outer solution is known to be the exact one for the quasi-static case of a stationary crack ($M=0$) with the coordinate of the point $E(x_E, 0)$ being $x_E=x_p/2$. We expect, therefore, that (4.20) will be at least approximately valid for small M.

For small values of ψ we have

$$w\cong\left(\frac{R}{2\pi}\right)^{1/2}\frac{2}{\mu}k_{\mathrm{III}}\frac{1}{2}\psi \tag{4.21}$$

$$\tau_{xz}\cong-\left(\frac{1}{2\pi R}\right)^{1/2}k_{\mathrm{III}}\frac{1}{2}\psi,\quad \tau_{yz}\cong\left(\frac{1}{2\pi R}\right)^{1/2}k_{\mathrm{III}}. \tag{4.22a,b}$$

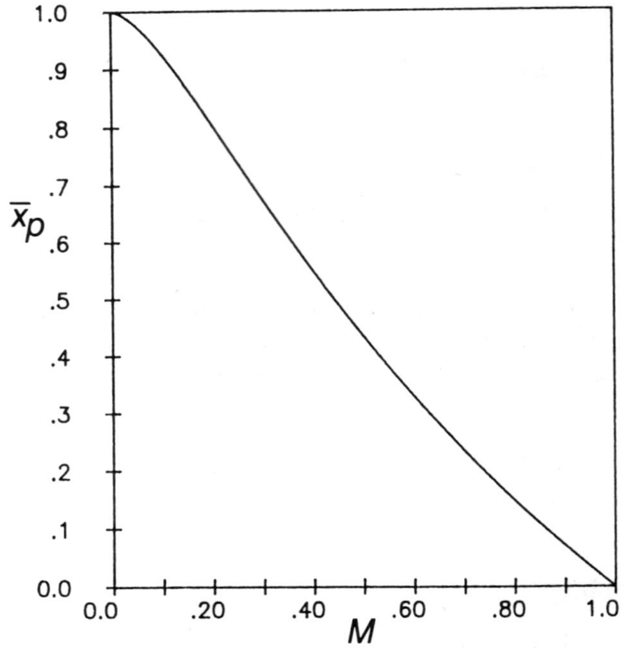

Figure 3. Position of transition boundary on crack-line, $\bar{x}_p = x_p\pi(k/k_{III})^2$, versus M.

Since τ_{yz} is continuous at $y=0$, $x=x_p$, see (4.13a), we find

$$\left(\frac{1}{2\pi R_p}\right)^{1/2} k_{III} = k, \quad \text{or} \quad R_p = (k_{III})^2/2\pi k^2, \tag{4.23a,b}$$

where $R_p = R|_{x=x_p}$. From (4.22a), (4.22b) and (4.23a) we find

$$\frac{\bar{\tau}_1}{k} = -\frac{1}{2}\frac{\psi}{y} \cong -\frac{1}{2R_p}, \quad \text{or} \quad \bar{\tau}_1 = -k/2R_p. \tag{4.24}$$

It then follows from (4.14a) that

$$\bar{\gamma}_1 = -k/2\mu R_p. \tag{4.25}$$

This result, together with (4.11), (4.12), (4.14a) and (4.16), yields

$$x_p = \frac{1}{\pi}\left(\frac{k_{III}}{k}\right)^2 \frac{1 - M^4 + 2M^2 \ln M}{(1-M^2)^2}. \tag{4.26}$$

A plot of $x_p\pi(k/k_{III})^2$ versus M is given in Fig. 3.

It is noted that x_p vanishes as $k \to \infty$. In that limit, the zone of nonlinear deformation shrinks onto the crack tip, and the linearly elastic solution represented by (4.21) and (4.22a,b) becomes the crack tip field.

5. Crack-line solutions for the elastic perfectly-plastic model

Analogously to (4.5), we consider expansions with respect to y along the crack line $y=0$, $0 < x \leqslant x_p$, where $x = x_p$ defines the elastic-plastic boundary

$$\tau_{yz} = s_0(x,t) + s_2(x,t)y^2 + O(y^4), \quad \tau_{xz} = \tau_1(x,t)y + O(y^3) \tag{5.1a,b}$$

$$\dot{w} = \dot{w}_1(x,t)y + O(y^3), \quad \dot{\Lambda} = \dot{\Lambda}_0(x,t) + \dot{\Lambda}_2(x,t)y^2 + O(y^4). \tag{5.2a,b}$$

Equations (5.1)–(5.2) take into account that τ_{yz} and $\dot{\Lambda}$ are symmetric with respect to $y=0$, while τ_{xz} and $\dot{w}$ are antisymmetric. Substitution of (5.1)–(5.2) into (3.7), (3.1) and (3.8a, b), and collecting terms of the lowest orders in y, yields

$$\frac{\partial \tau_1}{\partial x} + 2s_2 = \rho \ddot{w}_1 \tag{5.3}$$

$$s_0^2 = k^2, \quad 2s_0 s_2 + \tau_1^2 = 0 \tag{5.4a,b}$$

$$\frac{1}{2}\frac{\partial \dot{w}_1}{\partial x} = \frac{\dot{\tau}_1}{2\mu} + \dot{\Lambda}_0 \tau_1, \quad \tfrac{1}{2}\dot{w}_1 = \dot{\Lambda}_0 s_0. \tag{5.5a,b}$$

It follows from (5.4a) that $s_0 = k$. Elimination of s_2 from (5.3) and (5.4b) gives

$$\frac{\partial \tau_1}{\partial x} - \frac{\tau_1^2}{k} - \rho \ddot{w}_1 = 0. \tag{5.6}$$

Similarly, $\dot{\Lambda}_0$ can be eliminated from (5.5a) and (5.5b) to yield

$$\frac{\partial \dot{w}_1}{\partial x} - \frac{\dot{\tau}_1}{\mu} - \frac{1}{k}\dot{w}_1 \tau_1 = 0. \tag{5.7}$$

Equations (5.6) and (5.7) define two coupled nonlinear partial differential equations. Analytical solutions to these equations, which would give the transient fields on the crack line, have not yet been obtained.

Equations (5.6) and (5.7) must be supplemented by conditions at the elastic-plastic boundary Σ, see Fig. 1. It can be shown that near the crack line at $x = x_p$, we can write

$$[[s_0]] = 0, \quad [[\tau_1]] = 0 \quad \text{and} \quad [[\dot{w}_1]] = 0. \tag{5.8a,b,c}$$

The governing equations for the quasi-static case follow by setting $\rho \equiv 0$. The resulting system of coupled nonlinear ordinary differential equations can be solved. The quasi-static solution for $\dot{w}_1$ has been given in [5].

For the steady-state case at constant crack tip speed, the material time derivatives reduce to the form given by (4.1). Then $\dot{w}_1 = -\dot{a}\gamma_1$, where γ_1 is defined by (4.9). Equations (5.7) and (5.6) may then be written as

$$\frac{d\gamma_1}{dx} - \frac{1}{\mu}\frac{d\tau_1}{dx} - \frac{1}{k}\gamma_1 \tau_1 = 0 \tag{5.9}$$

$$\frac{d\tau_1}{dx} - \frac{\tau_1^2}{k} - \mu M^2 \frac{d\gamma_1}{dx} = 0, \tag{5.10}$$

where M is defined by (2.5).

It is of interest to note that an asymptotic solution for small values of x can be obained with minimal effort. Let us consider solutions of the general form $\tau_1 \cong -\tau_1^*/x$, $\gamma_1 \cong \gamma_1^*/x$. Substitution in (5.9) and (5.10) and collecting terms of $O(1/x^2)$ yields $\tau_1^* = k(1 \pm M)$, $\gamma_1^* = \pm(1 \pm M)k/\mu M$. Since we must have $\gamma_1 < 0$, we discard the solution containing the plus signs. Hence

$$\tau_1 \cong -\frac{k(1-M)}{x}, \quad \gamma_1 \cong -\frac{k}{\mu}\frac{1-M}{M}\frac{1}{x}. \tag{5.11a,b}$$

Since $\gamma_1 = \partial\gamma_y/\partial x$, we also have

$$\gamma_y \cong -\frac{k}{\mu}\frac{1-M}{M}\ln(x/x_p). \tag{5.12}$$

This solution is the same as the one derived earlier by Slepyan [8], see also Achenbach and Dunayevsky [9]. Note that τ_1 reduces to the quasi-static solution as $M \to 0$. The strain γ_y has, however, not only the wrong behavior in x, but it actually becomes singular in M.

Exact solutions to (5.9) and (5.10) can be obtained in implicit form. First we define the following dimensionless quantities:

$$T=\tau_1/\bar{\tau}_1, \quad \Gamma=-\mu\gamma_1/2\bar{\tau}_1, \quad X=-2\bar{\tau}_1 x/k, \tag{5.13a,b,c}$$

where $\bar{\tau}_1$ is defined by (4.15a). Equations (5.9) and (5.10) then become

$$2\frac{\mathrm{d}\Gamma}{\mathrm{d}X}+\frac{\mathrm{d}T}{\mathrm{d}X}+T\Gamma=0 \tag{5.14}$$

$$2\frac{\mathrm{d}T}{\mathrm{d}X}+T^2+4M^2\frac{\mathrm{d}\Gamma}{\mathrm{d}X}=0. \tag{5.15}$$

Next, consider

$$F=\frac{1}{T^2}\frac{\mathrm{d}T}{\mathrm{d}X}. \tag{5.16}$$

From (5.14)–(5.16) we then obtain

$$\Gamma=\frac{T}{M^2}\left[(1-M^2)F+\tfrac{1}{2}\right]. \tag{5.17}$$

Substitution of (5.17) into (5.15) yields by the use of (5.16)

$$2(1-M^2)\left(F\frac{\mathrm{d}T}{\mathrm{d}X}+T\frac{\mathrm{d}F}{\mathrm{d}X}\right)+2\frac{\mathrm{d}T}{\mathrm{d}X}+\frac{1}{2}T^2=0. \tag{5.18}$$

The solution to (5.18) is

$$T=C\frac{|2F+1/(1+M)|^{(1-M)/2M}}{|2F+1/(1-M)|^{(1+M)/2M}}, \tag{5.19}$$

where C is an integration constant.

Since $\bar{\tau}_1=\mu\bar{\gamma}_1$ at $x=x_p$, we have

$$\text{at } X=X_p=-2\bar{\tau}_1 x_p/k: \qquad \Gamma=-\tfrac{1}{2}, \quad T=1. \tag{5.20}$$

By the use of (5.20) we find from (5.17)

$$F|_{x=x_p}=-\frac{1}{2}\frac{1+M^2}{1-M^2}. \tag{5.21}$$

It then follows from (5.19) that

$$C=\left(\frac{1-M}{1+M}\right)^{1/2M}\frac{M}{(1-M^2)^{1/2}}. \tag{5.22}$$

Integration of (5.16) yields

$$X=\int_1^T\frac{\mathrm{d}T}{T^2F(T)}+X_p. \tag{5.23}$$

Since $T\to\infty$ as $X\to 0$, we obtain from (5.23)

$$X_p=-\int_1^\infty\frac{\mathrm{d}T}{T^2F(T)}. \tag{5.24}$$

For fixed M, (5.19) and (5.22) relate F and T. The function $F(T)$ can be obtained numerically. Substitution of the result in (5.23) yields $T(X)$. The latter result can be substituted in (5.17) to yield $\Gamma(X)$. Finally, $\gamma_1=\mathrm{d}w_1/\mathrm{d}x$ can be used to compute the strain on the crack line:

$$\gamma_y=\frac{k}{\mu}+\gamma_y^p=\frac{k}{\mu}+\int_{x_p}^{x}\gamma_1\mathrm{d}x. \tag{5.25}$$

Substitution of (5.13b, c) yields

$$\frac{\mu}{k}\gamma_y^p = \int_{X_p}^{X}\Gamma(X)\mathrm{d}X = X_p\int_1^{x/x_p}\Gamma(X_p\xi)\mathrm{d}\xi, \tag{5.26}$$

where X_p is given by (5.24).

Equation (5.26) is an *exact* result for the plastic strain versus the relative distance to the elastic-plastic boundary, irrespective of the external loading. Of course, the distance x_p does depend on the loading and the overall geometry. Equation (5.26) is equivalent to the solutions obtained by Dunayevsky and Achenbach [10] and Freund and Douglas [11]. For various values of M, the plastic strain has been plotted versus x/x_p in Fig. 4. It can be shown that for small x, (5.26) reduces to (5.12). In the limit $M \to 0$ (5.26) reduces to the quasi-static solution

$$\frac{\mu}{k}\gamma_y = 1 - \ln(x/x_p) + \tfrac{1}{2}\left[\ln(x/x_p)\right]^2, \tag{5.27}$$

provided that $|M \ln(x/x_p)| \ll 1$. It appears that there is a small zone near the crack tip in which γ_y is given by (5.12). As M decreases, this zone becomes smaller, shrinking to zero as $M \to 0$. Hence the apparent singular behavior with M of (5.11)–(5.12) is not real, because the zone of applicability of these expressions vanishes in the limit $M \to 0$.

An explicit expression for x_p in terms of M can be achieved by matching the field inside the plastic zone to an elastic field of the form (4.20). Substitution of the results $\bar{\tau}_1 = -k/2R_p$, (4.24), and (4.23b) into (5.20) yields $X_p = 2\pi(k/k_{III})^2 x_p$. This relation and (5.24), with $F(T)$ from (5.19), give x_p as a function of M. Figure 5 shows $2x_p\pi(k/k_{III})^2$ versus M, as computed from (5.24).

Finally, following Freund and Douglas [11] we apply the crack growth criterion of critical plastic strain to determine the value of k_{III} that would be required for crack growth at a given value of M. The crack-growth criterion, originally proposed by McClintock and Irwin [12], states that the crack will grow with (normalized) plastic strain $(\mu/k)\gamma_y^p = \gamma_f$ at $x = x_f$ on $y = 0$. For plastic strain below γ_f at $x = x_f$ the crack cannot grow. As discussed by Rice [13] the characteristic length x_f is related to K_c, the value of the mode III stress intensity factor which is required to satisfy the fracture criterion for a

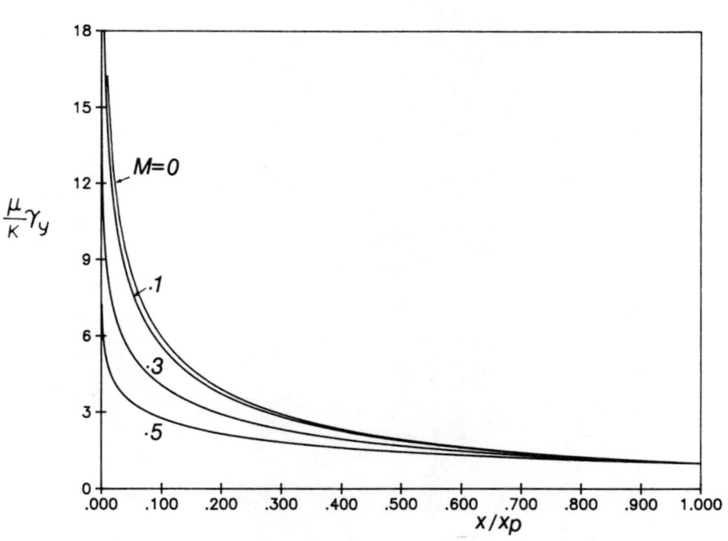

Figure 4. Strain, $(\mu/k)\gamma_y$, on crack line versus x/x_p for various values of M.

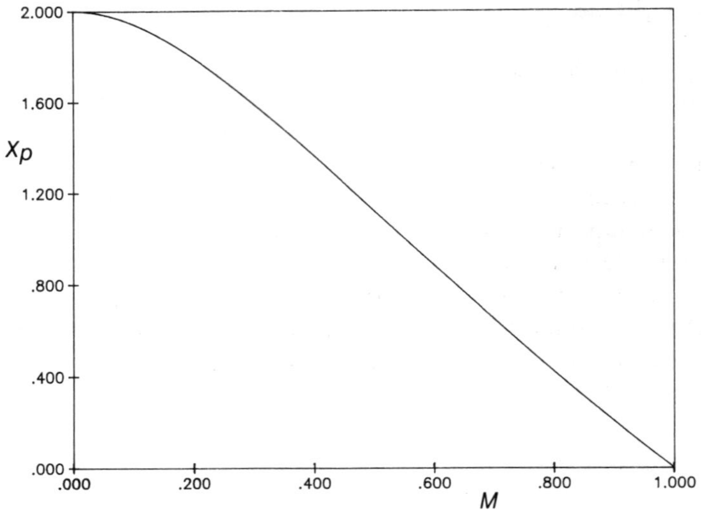

Figure 5. Position of elastic-plastic boundary on crack line, $X_p = 2\pi x_p(k/k_{III})^2$, versus M.

stationary crack, by the relation

$$\pi x_f(\gamma_f + 1) = (K_c/k)^2, \quad \text{or} \quad x_f = (K_c/k)^2/\pi(\gamma_f + 1). \tag{5.28}$$

We can now compute γ_y^p at $x = x_f$ from (5.26). The equality $\gamma_f = (\mu/k)\gamma_y^p(x_f/x_p)$, where x_p follows from (5.24), yields a relation between k_{III}/K_c and γ_f and M. For three values of γ_f, the relation between k_{III}/K_c and M has been plotted in Fig. 6. It is noted that the elastodynamic stress intensity factor k_{III} is the *dynamic* factor, which is related to the corresponding quasi-static factor by (2.4). Hence, in comparison with the quasi-static case, the remote load to attain a high crack tip speed is actually even higher.

The increase of k_{III}/k with M shown in Fig. 6 is sometimes interpreted as indicating an "increased fracture toughness" under fast fracture conditions. This is a misleading statement. In the present treatment, the material parameter is the critical strain γ_f. Figure 6 simply shows that for a certain γ_f a higher external load is required to achieve a higher crack tip speed.

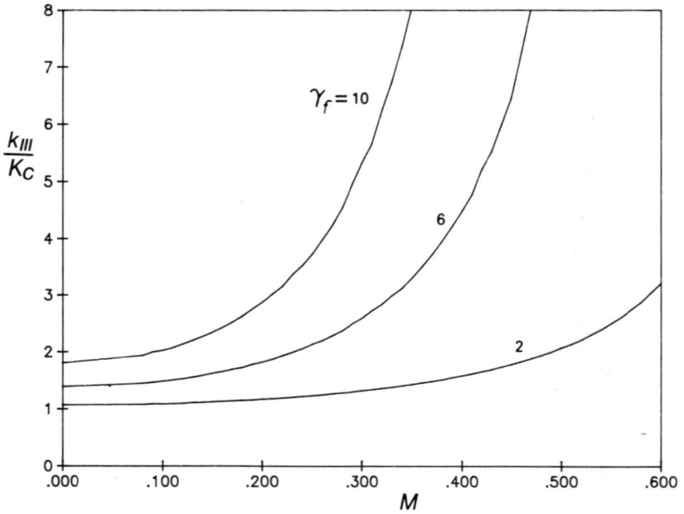

Figure 6. Ratio k_{III}/K_c versus M for various values of γ_f.

The method of constructing crack-line fields presented in this section has been extended in an approximation manner to dynamic crack propagation under plane stress conditions in an elastic perfectly-plastic material [14].

Other analytical work has been strictly limited to the asymptotic near-tip fields. The investigations of the asymptotic structure of the dynamic near-tip fields in elastic perfectly plastic materials by Slepyan [8] and Achenbach and Dunayevsky [9] have recently been reconsidered by Gao and Nemat-Nasser [15]. For the plane-strain case these authors have modified the results of [8] and [9] to include possible discontinuities. Recent numerical results by Lam and Freund [16] indicate, however, that discontinuities do not exist, and that the results of [8] and [9] are correct. Dynamic near-tip effects for a strain-hardening material were investigated by Achenbach and Kanninen [17] and Achenbach, Kanninen and Popelar [18] on the basis of J_2-flow theory and a bilinear effective stress-strain relation.

Acknowledgment

This paper was written in the course of research sponsored by the Office of Naval Research (Contract No. N00014-76-C-0063) and the Air Force Office of Scientific Research (Grant AFOSR-83-0308).

References

[1] J.D. Achenbach, in *Mechanics Today*, edited by S. Nemat-Nasser, 1, Pergamon Press (1974) 1.
[2] L.B. Freund, in *The Mechanics of Fracture,* edited by F. Erdogan, ASME-AMD 19 (1975) 105.
[3] M.F. Kanninen, in *Numerical Methods in Fracture Mechanics,* edited by A.R. Luxmoore and D.R.J. Owen, University College of Swansea Press (1978) 612.
[4] J.D. Achenbach and Z.P. Bazant, *Journal of Applied Mechanics* 42 (1975) 183.
[5] J.D. Achenbach and V. Dunayevsky, *Journal of the Mechanics and Physics of Solids* 32 (1984) 89.
[6] J.D. Achenbach and Z.L. Li, *Engineering Fracture Mechanics* (1985), in press.
[7] J.D. Achenbach and N. Nishimura, *Journal of Applied Mechanics,* to appear.
[8] L.I. Slepyan, *Izv. Nauk SSSR, Mekhanika Tverdogo Tela* 12 (1976) 144.
[9] J.D. Achenbach and V. Dunayevsky, *Journal of the Mechanics and Physics of Solids* 29 (1981) 283.
[10] V. Dunayevsky and J.D. Achenbach, *International Journal of Solids and Structures* 18 (1982) 1.
[11] L.B. Freund and A.S. Douglas, *Journal of the Mechanics and Physics of Solids* 30 (1982) 59.
[12] F.A. McClintock and G.R. Irwin, in *Fracture Toughness and Testing and its Applications,* ASTM-STP 381 (1965).
[13] J.R. Rice, in *Fracture II*, edited by H. Liebowitz, Academic Press (1968) 191.
[14] J.D. Achenbach and Z.L. Li, in *Fundamentals of Deformation and Fracture*, edited by B.A. Bilby and K.J. Miller, Cambridge University Press, to appear.
[15] Y.C. Gao and S. Nemat-Nasser, *Mechanics of Materials* 2 (1983) 47.
[16] P.S. Lam and L.B. Freund, *Journal of the Mechanics and Physics of Solids,* to appear.
[17] J.D. Achenbach and M.F. Kanninen, in *Fracture Mechanics*, edited by N. Perrone et al., University of Virginia Press (1978) 649.
[18] J.D. Achenbach, M.F. Kanninen and C.H. Popelar, *Journal of the Mechanics and Physics of Solids* 29 (1981) 211.

International Journal of Fracture 27 (1985) 229–243.

Computational methods based on an energy integral in dynamic fracture

T. NAKAMURA, C.F. SHIH and L.B. FREUND
Division of Engineering, Brown University, Providence, RI 02912, USA

(Received September 1, 1984)

Abstract

Developments in the use of the crack tip energy flux integral in computational dynamic fracture mechanics over the past few years are reviewed. An expression for the crack tip energy flux in terms of near tip mechanical fields which is valid for general material response is derived. It is then demonstrated that certain useful energy integrals may be extracted from the general result by invoking the appropriate characterization of material response. Several alternative representations of energy flux in the form of integrals over some finite region around the crack tip are presented and compared with a view toward implementation in finite element simulation studies.

1. Introduction

The first crack tip contour integral expression for elastodynamic energy release rate was proposed by Atkinson and Eshelby [1], who argued that the form for dynamic growth should be the same as for quasi-static growth with the elastic energy density replaced by the total mechanical energy density, that is, the elastic energy plus the kinetic energy. The equivalent integral expression for dynamic energy release rate in terms of crack tip stress and deformation fields was subsequently derived directly from the field equations of elastodynamics by Kostrov and Nikitin [2] and by Freund [3]. They enforced an instantaneous energy rate balance for the time-dependent volume of material bounded by the outer boundary of the solid, the crack faces, and small loops surrounding each moving crack tip and translating with it. By application of Reynold's transport theorem [4] and the divergence theorem an expression for crack tip energy flux in the form of an integral of field quantities along the crack tip loop was obtained. A result which is applicable to a much broader range of material response and which contains the earlier elastodynamic result as a special case is derived here, with an observation made by Eshelby [5] on the form of the equations of motion as a point of departure. A similar result has also been obtained by Willis [6].

In the analysis of dynamic crack phenomena by means of computational methods, such as the finite element method, a fundamental difficulty is encountered in efforts to compute values of the crack tip energy flux versus time or amount of crack growth. The difficulty arises from the fact that, on the one hand, the crack tip energy flux is defined in terms of the values of field quantities for points arbitrarily close to the crack tip while, on the other hand, it is precisely for points near the crack tip for which the accurate calculation of field quantities is most difficult. In an effort to circumvent this difficulty, finite domain integrals have been introduced for the computation of dynamic energy release rate by Kishimoto, Aoki and Sakata [7,8], by Atluri [9], and by Nishioka and Atluri [10–12].

The purpose here is to review developments in this area over the past few years. First, an expression for crack tip energy flux which is valid for general material response is derived in a direct manner. While a variety of seemingly different path-independent energy integrals have been proposed in recent years, each may be extracted from this general result by invoking the appropriate restrictions on material response and crack tip motion. Several alternate expressions for the energy flux integral are obtained from the general result by application of the divergence theorem over some part of the fracturing solid and these expressions are compared, with a view toward implementation in finite element simulations. For the illustrations based on elastodynamic crack growth, exact solutions are available to test the computed quantities. In the case of elastic-plastic response, on the other hand, no exact solutions exist to provide a basis for comparison.

2. The crack tip energy flux integral

In the absence of body forces, the equation of motion in terms of cartesian components of stress σ_{ij} and displacement u_i is

$$\frac{\partial \sigma_{ij}}{\partial x_j} - \rho \frac{\partial^2 u_i}{\partial t^2} = 0, \tag{2.1}$$

where ρ is mass density. If the inner product of (2.1) with the particle velocity $\partial u_i/\partial t$ is formed, then the resulting equation may be written in the form

$$\frac{\partial}{\partial x_j}\left(\frac{\partial u_i}{\partial t}\sigma_{ij}\right) - \frac{\partial}{\partial t}(U+T) = 0, \tag{2.2}$$

where U is the stress work density and T is the kinetic energy density, namely, at any material point,

$$U = \int_0^t \sigma_{ij} \frac{\partial^2 u_i}{\partial t' \partial x_j} \mathrm{d}t' \qquad T = \tfrac{1}{2}\rho \frac{\partial u_i}{\partial t}\frac{\partial u_i}{\partial t}. \tag{2.3}$$

Equation (2.2) is a differential form of the work balance principle which is valid for any material response.

If the time coordinate is now given the same status as a spatial coordinate, then (2.2) is the requirement that a certain vector expression is divergence free in a particular "volume" of space-time, that is, it represents a mechanical conservation law. Thus, if the divergence theorem is applied over this volume, the result is that the integral over the bounding surface of this volume of the inner product of the divergence free vector with the surface unit normal vector vanishes. For definiteness, consider a two-dimensional body which contains an extending crack and bounded by a curve C (attached to arbitrary material particles). The plane of the body is the x_1, x_2-plane, the crack plane is $x_2 = 0$ and crack growth is along the x_1-axis. The instantaneous crack tip speed is v. A small contour Γ begins on one traction free face of the crack, surrounds the tip, and ends on the opposite traction free face. This crack tip contour is fixed in size and orientation with respect to the crack tip coordinates as the crack grows. Consider the volume in the x_1, x_2, t-space that is bounded by the planes $t = t_1$ and $t = t_2$, where t_1 and t_2 are arbitrary times, by the right cylinder swept out by C between the two times, by the planes swept out by the crack faces between the two times, and by the tubular surface swept out by the small contour Γ as the crack advances in the x_1-direction with increasing time. For this particular choice of surface, the surface integral which results from application of the divergence theorem and Reynold's transport theorem is

$$\int_{t_1}^{t_2}\int_C \sigma_{ij}n_j\frac{\partial u_i}{\partial t}\mathrm{d}C\mathrm{d}t=\left[\int_{A(t)}(U+T)\mathrm{d}A\right]_{t_1}^{t_2}+\int_{t_1}^{t_2}\int_\Gamma\left(\sigma_{ij}n_j\frac{\partial u_i}{\partial t}+(U+T)vn_1\right)\mathrm{d}\Gamma\mathrm{d}t, \tag{2.4}$$

where $A(t)$ is the cross-section of the above-mentioned volume on any plane $t=$ constant, n_i is a unit vector that is normal to C or Γ in any plane $t=$ constant and that points away from the crack tip, and vn_1 is the instantaneous velocity of any point on Γ in the direction of n_i at this point.

The term on the left side of (2.4) is the total work done on the body in the time interval of interest, and the first term on the right side is the increase in internal energy during the same time. Consequently, the last term in (2.4) is the total energy lost from the body due to flux through Γ during the interval $t_1 \le t \le t_2$. Because this time interval is completely arbitrary, it follows immediately that the crack tip line integral

$$F(\Gamma)=\int_\Gamma\left(\sigma_{ij}n_j\frac{\partial u_i}{\partial t}+(U+T)vn_1\right)\mathrm{d}\Gamma \tag{2.5}$$

is the instantaneous rate of energy flow out of the body through Γ. The interpretation of the terms in (2.5) is straightforward. The first term is the rate of work of the material outside of Γ on the material inside of Γ. If the curve were a material line, then this would be the only contribution to the energy flux. The curve is moving through the material, however, and the second term in (2.5) represents the contribution to the energy flux due to the flux of material.

The dynamic energy release rate $\mathcal{G}$, that is, the energy released from the body per unit crack advance, is defined as the limiting value of $F(\Gamma)/v$ as the contour Γ is shrunk onto the crack tip. In order for this concept to have any fundamental significance, it is necessary that the result of the limiting process be independent of the actual shape of Γ as its size becomes vanishingly small. Surely F itself is not path-independent in general, as can be seen by means of a direct check. Consider the closed path formed by two crack tip contours Γ_1 and Γ_2, and by the segments of the crack faces that connect the ends of these contours. Application of the divergence theorem to the energy flux integral for this entire closed path leads to the result that

$$F(\Gamma_2)-F(\Gamma_1)=\int_{A_{12}}\left[\left(\frac{\partial U}{\partial t}+v\frac{\partial U}{\partial x_1}\right)+\rho\frac{\partial u_i}{\partial t}\left(\frac{\partial^2 u_i}{\partial t^2}+v\frac{\partial^2 u_i}{\partial t\partial x_1}\right)\right]\mathrm{d}A, \tag{2.6}$$

where A_{12} is the area within the closed path. An equivalent expression for linearly elastic response appears as (14) in [3]. For points very close to the crack tip, it is evident that F will indeed be path-independent (provided that the singularities in field quantities are such that the integral defining F exists near the crack tip) in the limit as Γ shrinks onto the tip if u_i is smooth enough for its second derivatives to exist and if

$$\frac{\partial f}{\partial t}+v\frac{\partial f}{\partial x_1}=o\left(\frac{\partial f}{\partial x_1}\right) \quad \text{as} \quad r\to 0^+, \tag{2.7}$$

where f is any field variable and r is the radial distance from the moving crack tip. It is noted that the right side of (2.6) is zero by definition for steady-state crack growth problem formulations, so that (2.5) provides a path-independent integral for this class of problems for any type of material response.

If (2.7) is assumed to hold for arbitrary crack tip motion under dynamic conditions, then $\partial u_i/\partial t$ may be replaced by $-v\partial u_i/\partial x_1$ in (2.5) so that

$$\mathcal{G}=\lim_{\Gamma\to 0}\int_\Gamma\left((U+T)n_1-\sigma_{ij}n_j\frac{\partial u_i}{\partial x_1}\right)\mathrm{d}\Gamma. \tag{2.8}$$

Any difference between $\partial u_i/\partial t$ and $-v\partial u_i/\partial x_1$ which is less singular than $\partial u_i/\partial x_1$ cannot contribute to the value of $\mathscr{G}$. The general result (2.8) underlies virtually all crack tip energy integrals that have been defined and applied in fracture mechanics, in the sense that each may be derived from (2.8) by invoking the appropriate restrictions on material response (through U) and on crack tip motion. Some illustrations and applications are cited in the following discussion, with particular emphasis on the computation of crack tip energy quantities.

3. Elastodynamics

3.1. Linearly elastic materials

For linearly elastic material response, $U=\frac{1}{2}\sigma_{ij}\partial u_i/\partial x_j$ and the stress components have the characteristic inverse square root dependence on radial distance from the crack tip. The dynamic energy release rate $\mathscr{G}$ is proportional to the square of the coefficient of this square root singularity, the so-called elastic stress intensity factor. For a growing crack, the particle velocity also has the inverse square root dependence, whereas the particle velocity vanishes as the square root of the distance from the crack tip for a stationary crack under dynamic loading. The case of dynamic crack growth is considered first.

Consider an arbitrary material area A_o around the crack tip bounded by the curve Γ_o. The curve Γ_o begins on one crack face and ends on the opposite face, as shown in Fig. 1. To be somewhat more precise, A_o is the area enclosed by Γ_o, Γ and the crack faces in the limit as Γ is shrunk onto the crack tip. Application of the divergence theorem to (2.8) leads immediately to the expression,

$$\mathscr{G}=\int_{\Gamma_o}\left(Un_1-\sigma_{ij}n_j\frac{\partial u_i}{\partial x_1}+\tfrac{1}{2}\rho v^2\frac{\partial u_i}{\partial x_1}\frac{\partial u_i}{\partial x_1}n_1\right)\mathrm{d}\Gamma$$

$$+\int_{A_o}\left(\rho\frac{\partial^2 u_i}{\partial t^2}\frac{\partial u_i}{\partial x_1}-\rho v^2\frac{\partial u_i}{\partial x_1}\frac{\partial^2 u_i}{\partial x_1^2}\right)\mathrm{d}A. \tag{3.1a}$$

At first sight, it might appear that the area integral in (3.1a) is divergent because $\partial u_i/\partial x_1$ and $\partial^2 u_i/\partial t^2$ are singular as $r^{-1/2}$ and $r^{-3/2}$, respectively. This is not so, however, due to the particular combination of terms in the integrand. The displacement gradient near the crack tip has the form $\partial u_i/\partial x_1 = r^{-1/2}f_i(\theta)$ in crack tip polar coordinates, so that the

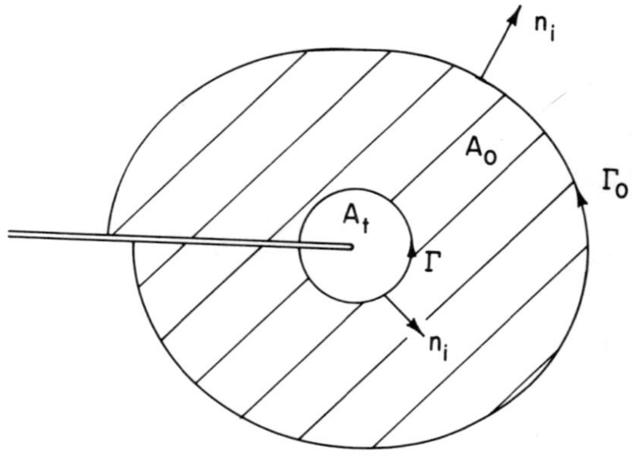

Figure 1. Domain A_o in x_1-x_2 plane is enclosed by Γ, Γ_o and crack faces. A_t is crack tip region enclosed by Γ and crack faces.

second derivative has the form $\partial^2 u_i/\partial x_1^2 = r^{-3/2} g_i(\theta) = -r^{-3/2}[\frac{1}{2} f_i(\theta)\cos\theta + f_i'(\theta)\sin\theta]$. For any angular variation $f_i(\theta)$, it is easily shown that

$$\int_{-\pi}^{\pi} f_i(\theta) g_i(\theta) \mathrm{d}\theta = 0 \tag{3.2}$$

provided that $f_i(\pm\pi)$ is bounded. Thus, the nonintegrable singularity in the area integral in (3.1a) is only apparent and the integral does indeed have a finite value.

While the representation for $\mathcal{G}$ in (3.1a) is certainly correct, it has a feature which is the basis for some concern in considering implementation in finite element procedures. The second term of the integrand in the area integral contains the derivatives of displacement gradients. To obtain non-vanishing second derivatives of the displacements, the elements must have quadratic or higher order interpolation functions for the displacements. Thus, certain classes of elements, including the commonly employed linear displacement elements, are ruled out for such calculations. Furthermore, we note that the products of stress and displacement gradients appear naturally in the virtual work statement of equilibrium and such products can be calculated accurately at the Gauss integration points. However, this accuracy is not preserved for the higher derivatives. Hence, representation (3.1a) may not be ideal for computing the energy release rate.

Other representations for energy release rate may be obtained as follows,

$$\mathcal{G} = \int_{\Gamma_o}\left(Un_1 - \sigma_{ij}n_j\frac{\partial u_i}{\partial x_1} + \tfrac{1}{2}\rho\frac{\partial u_i}{\partial t}\frac{\partial u_i}{\partial t}n_1\right)\mathrm{d}\Gamma + \int_{A_o}\left(\rho\frac{\partial^2 u_i}{\partial t^2}\frac{\partial u_i}{\partial x_1} - \rho\frac{\partial u_i}{\partial t}\frac{\partial^2 u_i}{\partial x_1 \partial t}\right)\mathrm{d}A \tag{3.1b}$$

$$\mathcal{G} = \lim_{\Gamma\to 0}\int_{\Gamma}\tfrac{1}{2}\rho v^2\frac{\partial u_i}{\partial x_1}\frac{\partial u_i}{\partial x_1}n_1\mathrm{d}\Gamma + \int_{\Gamma_o}\left(Un_1 - \sigma_{ij}n_j\frac{\partial u_i}{\partial x_1}\right)\mathrm{d}\Gamma + \int_{A_o}\rho\frac{\partial^2 u_i}{\partial t^2}\frac{\partial u_i}{\partial x_1}\mathrm{d}A. \tag{3.1c}$$

In (3.1c) the first integral is evaluated at the crack tip and the remaining two integrals are evaluated over the finite region. As noted earlier, an accurate evaluation of the integral at the crack tip is virtually impossible to achieve. However, if the contribution from the first integral is small relative to contributions from the other integrals, then (3.1c) could be a useful representation. To pursue this point, note that for linearly elastic fields, the first integral on the right in (3.1c) can be evaluated using the known crack tip field quantities [13] to yield a result which is proportional to the stress intensity factor squared and which depends on the instantaneous crack tip speed. If each term in (3.1c) is divided by $\mathcal{G}$, which is also proportional to the stress intensity factor squared and which depends on the crack tip speed, then the first term on the right in the resulting equation is a function of crack tip speed only. If this speed dependent ratio (denoted by R below) is much smaller than unity, then the first term can be neglected and a relatively simple expression for energy release rate is obtained.

Choosing Γ to be a circular path and using the plane strain relationship between the energy release rate $\mathcal{G}$ and stress intensity factor K given in [13], the ratio R is given by

$$R(v) = \left|\frac{\int_{-\pi}^{\pi}\tfrac{1}{2}\rho v^2\frac{\partial u_i}{\partial x_1}\frac{\partial u_i}{\partial x_1}n_1 r\mathrm{d}\theta}{(1-\nu^2)A(v)K^2/E}\right|, \tag{3.3}$$

where ν is Poisson's ratio, E is the elastic modulus, K is the elastic stress intensity factor, and the dimensionless function of speed $A(v)$ is given in [13]. Values of this ratio have been computed using the explicit expressions for angular variation of crack tip fields given in [13]. From the results it is clear that for crack tip speeds smaller than half the shear wave speed, the integral contributes less than 10 percent to the energy release rate, however, the contribution becomes significant at higher crack tip speeds.

With regard to the observation made above, a feature of (3.1b) which is of importance from the point of view of computation is noted. The third and fifth terms in (3.1b) are typically comparable in magnitude but are of the opposite signs, and their net contribution to the energy release rate (i.e. the sum of the two terms) is small. In fact these terms are equivalent to the first term in (3.1c) which has been shown to contribute negligibly to the energy release rate at lower crack tip speed. In other words, a small contribution to the energy release rate is computed from the difference of the two terms of comparable magnitude. This is a potential source of computational error. We conclude that the difference between (3.1b) and (3.1c) is bounded and small for $v/c_2 < 0.5$. The result (3.2) also suggests that the relative contribution from the vicinity of the crack tip to the area integral in (3.1c) is small for uniformly rising loads. Our numerical calculations confirmed this behavior. Hence expression (3.1c) appears to be suited for evaluating the energy release rate from finite element field quantities for crack speeds $v/c_2 < 0.5$.

Finally, it is noted that the expression (3.1c) with $v \rightarrow 0$ is exactly the energy release rate expression for the case of a stationary crack subjected to dynamic loads. In this case, the quantity $\mathscr{G}$ must be interpreted as a virtual energy release rate, that is, $\mathscr{G}$ computed from (3.1c) with $v = 0$ is the energy change that would be released if the mechanical fields were frozen at any instant and the crack tip were given an increment of extension in the x_1-direction. Of course, Irwin's relationship between the stress intensity factor and $\mathscr{G}$ holds in this case, and the interpretation of $\mathscr{G}$ as a measure of crack tip field intensity rather than as an energy release rate is more attractive in some respects.

3.2. Nonlinearly elastic materials

All of the energy release rate expressions derived in the preceding section apply for nonlinearly elastic material response, as well as linearly elastic response. However, there are too few theoretical results available for dynamic fracture of nonlinearly elastic materials at the present time to provide a basis for a general discussion. For a stationary crack in a material with a power law stress-strain relationship, it would seem that the near tip fields are adequately described by the HRR asymptotic solution [14,15] provided that their time variation is not too rapid. Analysis of such problems would proceed as in the analysis of certain plasticity problems as considered below in Section 4.

On the other hand, it can be readily shown that a similar separable crack tip asymptotic field cannot be constructed for dynamic growth of a crack in the same material. An indication that analysis of dynamic crack growth in nonlinearly elastic materials may provide some surprises is contained in the recent work reported by Achenbach and Nishimura [16]. They considered the effect of inertia on mode III crack growth in a nonlinearly elastic material for which the effective stress has a constant magnitude for all effective strains greater than a particular level. A principal result is that the total strain remains bounded as the crack tip is approached from a direction directly ahead of the tip. This result poses a dilemma, for it suggests that there may be no energy released at the crack tip although energy continues to flow into the crack tip region from remote points. The resolution of this dilemma may be found through the observation that the material considered admits lines along which displacement gradients are discontinuous [17], and the energy flowing into the crack tip region may be dissipated through propagation of such lines of discontinuity.

4. Elastic-plastic material response

For a stationary crack in a dynamically loaded body of incremental elastic-plastic material, the integral quantity in (2.8) can still be used to some advantage. Of course, the interpretation of the integral as an energy release rate is lost for this case. Nonetheless, it has been demonstrated that if the loading applied to the crack tip region is monotonically increasing and if the stress histories for material particles near the crack tip are nearly proportional, then the value of the integral provides a measure of the intensity of the near tip deformation. It is this role that is pertinent to elastic-plastic fracture mechanics and considerable analytical, numerical and experimental investigations have shown that the values of this integral is a relevant characterizing parameter of the near tip fields of stationary cracks when certain size requirements are met. For dynamic loading conditions and a stationary crack tip, the kinetic energy term T makes no contribution to the integral in (2.8) so that

$$\mathscr{G} = J = \lim_{\Gamma \to 0} \int_{\Gamma} \left(U n_1 - \sigma_{ij} n_j \frac{\partial u_i}{\partial x_1} \right) \mathrm{d}\Gamma, \tag{4.1}$$

where U is a stress work density appropriate for proportional loading of the material. For quasistatic loading conditions and proportional stress history at each material point, the integral in (4.1) is path independent and its value does not depend on a limiting process in which the crack tip contour Γ is shrunk onto the crack tip. In this case, the integral is precisely Rice's J-integral [18], as anticipated in the notation in (4.1). The background for this approach to elastic-plastic fracture mechanics is outlined in recent review articles by Rice [19] and by Hutchinson [20]. In the case of dynamic near tip fields, on the other hand, inclusion of the limit in the definition of J is essential, that is, the J-integral is not path independent under dynamic loading conditions.

Just as in the case of analysis of elastic crack problems by means of computational techniques, the accurate determination of field quantities very near the crack tip is difficult and (4.1) cannot be evaluated directly. Application of the divergence theorem, however, leads to the alternate form

$$J = \int_{\Gamma_o} \left(U n_1 - \sigma_{ij} n_j \frac{\partial u_i}{\partial x_1} \right) \mathrm{d}\Gamma + \int_{A_o} \rho \frac{\partial^2 u_i}{\partial t^2} \frac{\partial u_i}{\partial x_1} \mathrm{d}A, \tag{4.2}$$

where the path Γ_o and the area A_o are as shown in Fig. 1. An illustration of the form (4.2) will be presented in the following section. Even in the case of dynamic growth of a crack in an incremental elastic-plastic material, the integral (2.8) can be applied for an appropriate definition of the term U. It has not been found to be particularly useful, however, because the near tip fields for growing cracks tend to be less singular than the corresponding fields for stationary cracks and the value of $\mathscr{G}$ is then zero. The physical interpretation of this result is that all of the energy that flows into the crack tip region is either dissipated through plastic flow or is trapped as residual elastic energy in the wake of the crack tip plastic zone.

An exception to this situation arises for rate-dependent plastic response when the rate sensitivity is sufficiently pronounced. A problem of a crack growing at high speed in an elastic/rate-dependent plastic solid under small scale yielding conditions was considered by Freund and Hutchinson [21]. At high stress levels, such as those that arise near the tip at high crack speed, it was assumed that the plastic strain rate increases linearly with stress. In this case, the near tip fields are sufficiently singular to result in a finite value for $\mathscr{G}$ in (2.8), and a general expression is derived in [21] for the near tip energy release rate in terms of the overall release rate for steady growth. In fact, for steady state growth, the integral in (2.8) is path independent for general material response [6,21].

5. Finite element analyses of stationary and propagating cracks

5.1. Elastodynamic analysis

Computational aspects of the domain integral representations for the energy release rate as given in Section 3 are discussed in this section. To emphasize the interpretation of these energy integrals (3.1) as a measure of the crack tip field intensity, the symbol J will be employed in place of $\mathcal{G}$ hereafter. In (3.1) the domain A_o is enclosed by Γ_o, Γ and the crack faces in the limit as Γ is shrunk onto the crack tip. From result (3.2), it is clear that the contributions from the area integrals in (3.1) taken over the domain A_t (indicated in Fig. 1) as Γ is shrunk onto the crack tip must vanish. Hence in the context of numerical integration A_o will be taken to be the entire area enclosed by Γ_o and the crack faces. The energy integrals have the form given in (3.1), of course.

The numerical studies were carried out with the finite element program ABAQUS [22]. The Hilber–Hughes–Taylor algorithm [23] is employed to integrate the semi-discrete equation of motion. A small amount of damping is invoked by setting the α value to be -0.05. Both 4-noded bilinear isoparametric element and 8-noded biquadratic isoparametric element are used in the calculations. Unless otherwise stated, element stiffness matrices are obtained by 2×2 Gaussian integrations.

To indicate the implementation of these energy integrals in our finite element calculations, the discrete form of (3.1b) is given as follows

$$J = \sum_{\text{all segments on } \Gamma_o} \sum_{i=1}^{2} w_i \left[\left(U - \sigma_{11}\frac{\partial u_1}{\partial x_1} - \sigma_{21}\frac{\partial u_2}{\partial x_1} + \tfrac{1}{2}\rho\frac{\partial u_1}{\partial t}\frac{\partial u_1}{\partial t} + \tfrac{1}{2}\rho\frac{\partial u_2}{\partial t}\frac{\partial u_2}{\partial t} \right) P_2 + \left(\sigma_{12}\frac{\partial u_1}{\partial x_1} + \sigma_{22}\frac{\partial u_2}{\partial x_1} \right) P_1 \right]_i$$
$$+ \sum_{\text{all elements in } A_o} \sum_{i=1}^{4} w_i \left[\left(\rho\frac{\partial^2 u_1}{\partial t^2}\frac{\partial u_1}{\partial x_1} + \rho\frac{\partial^2 u_2}{\partial t^2}\frac{\partial u_2}{\partial x_1} - \rho\frac{\partial u_1}{\partial t}\frac{\partial^2 u_1}{\partial x_1 \partial t} - \rho\frac{\partial u_2}{\partial t}\frac{\partial^2 u_2}{\partial x_1 \partial t} \right) D \right]_i . \tag{5.1}$$

Here P_1 and P_2 are the path factors given in [24], w_i are the appropriate Gauss weights and all quantities within $[\ ]_i$ are evaluated at the respective Gauss points. Also D is the determinant of the Jacobian matrix of the isoparametric transformation. Similar discrete expressions can be written for (3.1a) and (3.1c). A more detailed discussion of a discrete representation for a domain integral representation for the dynamic J can be found in [25].

The dynamic J-integrals (3.1) are independent of the size of the domain A_o enclosed by the contour Γ_o as indicated in Fig. 1. This invariant feature of the domain integrals can serve as a useful check on the self-consistency of the numerical calculations. As noted previously, accurate finite element solutions for points near the crack tip are difficult to obtain. Hence the domains of integrations typically extend over ten or more elements.

To evaluate the computational aspects of the numerical algorithm and of the energy integrals, a stress wave problem with an exact solution is employed as the benchmark problem. Freund [26] considered a semi-infinite crack in an infinite plane subjected to a tensile stress wave normal to the crack plane. At $t = 0$, the stress wave arrives at the crack plane and at a later time, $t = \tau$, the crack starts to propagate. The analytical solution for the dynamic stress intensity factor in the interval $0 < t < \tau$ (i.e., the crack is stationary) is

$$K(t) = B\sigma_a\sqrt{c_1 t}\,. \tag{5.2}$$

Here c_1 is the longitudinal wave speed for the medium, σ_a is the magnitude of the tensile pulse with step function time dependence and B is a function of Poisson's ratio given in [26]. For times $t > \tau$, the dynamic stress intensity factor for the propagating crack is

$$K(t, v) = k(v) B\sigma_a\sqrt{c_1 t}\,, \tag{5.3}$$

where v is the crack tip speed and $k(v)$ is defined in [27] as a function which decreases monotonically from one at $v = 0$ to zero when v is the Rayleigh wave speed.

For the purpose of comparison with numerical results presented subsequently, the well known relation between the dynamic J and the dynamic stress intensity factor under plane strain condition is given by

$$J(t, v) = \frac{(1-\nu^2)}{E} A(v)[K(t, v)]^2. \tag{5.4}$$

Here $A(v)$ is a universal function of crack tip speed [13]. For a stationary crack ($v = 0$), A is unity, and A increases monotonically with increasing v. In the subsequent analyses, ν is taken to be 0.3.

5.2. *Elastic analysis of a stationary crack subject to a single step pulse*

A step pulse is applied at $t = -H/c_1$ at the edge $x_2 = H$ of the model shown by inset in Fig. 2, and the pulse arrives at the crack plane at $t = 0$. Since the problem has no reflective symmetry with respect to the crack plane ($x_2 = 0$ plane), the domain downstream from the crack plane is included in the model. To simulate the infinite plane problem for which the solution (5.2) is applicable, the dimension H must be large enough so that within the period of interest, reflected waves from the boundaries do not reach the vicinity of the crack tip. In addition the length of the model (in x_1-direction) must be sufficiently large compared to H so that within the interval of interest, the crack tip region is unaffected by the unloading waves emanating from the remote edges. These factors were taken into consideration in selecting the dimensions of the finite element model. The full plane model constructed with 8-noded elements has 200 elements (10 by 20 arrangement) and 681

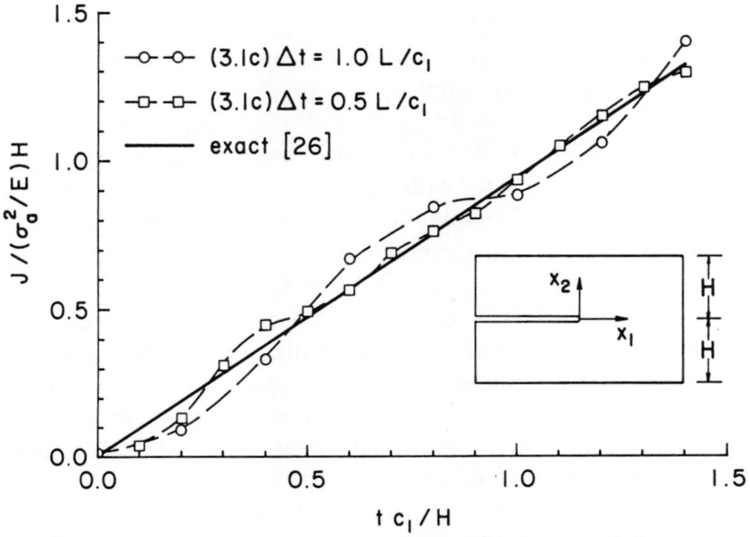

Figure 2. Variation of normalized J with normalized time as determined from the full plane model. Results obtained from two analyses using different time steps and the exact solution are shown. Shown in inset is a schematic of the full plane model and the fixed reference coordinates x_1 and x_2.

nodes. All elements are square shaped and are identical in size with the length (and width) being L. The time required for a longitudinal wave to traverse an element is then L/c_1.

Two calculations were carried out; the first employs uniform time steps of $\Delta t = 1.0L/c_1$ and the second employs uniform time steps of $\Delta t = 0.5L/c_1$. At the end of each time step, the dynamic J is computed from the stress and deformation fields using the discrete representation of (3.1c). Typically the dynamic J is evaluated on ten integration domains; the smallest domain has 4 elements and the largest domain has 200 elements. With the exception of the near-tip domain, which has four elements, the values of J evaluated on the other domains are within one percent of its mean value which attest to the consistency of the numerical solution. The dynamic J values determined from an alternate representation (3.1b) are also in agreement with the values determined from (3.1c). This provides an additional check on the consistency of the results. In subsequent discussions, the mean of the domain values will be reported as the dynamic J.

The dynamic J computed from (3.1c) is shown in Fig. 2 together with the exact solution (5.2) expressed in terms of J through (5.4). The computational solution clearly converges to the exact solution for the smaller time step. At sufficiently large times, say $t > 0.5H/c_1$, the finite element results for the smaller time step is within three percent of the exact solution. At shorter times, the error is substantially larger. This is not unexpected since at short times the refracted wave from the crack tip has traveled through relatively few elements (i.e., small domain), and as previously discussed, an accurate evaluation of the dynamic J in this case is rather difficult.

Similar calculations were also carried out using a mesh where the crack tip region is comprised of regular triangular and singular triangular elements [28]. Solutions from these models and from a model comprised solely of 4-noded elements (the latter model has 800 elements) are in substantial agreement with the numerical results shown in Fig. 2. Solutions from these models are also exhibit the same convergence characteristics described above. It is clear from this numerical study that (3.1c) and (3.1b) are suited for the determination of the dynamic J using field quantities obtained from finite element analysis.

5.3. Elastic analysis of a stationary and propagating crack subject to symmetrically applied step pulses

A computational analysis of crack propagation requires a finite element model which is substantially larger than the models employed in the preceding section for the stationary crack analysis. To reduce the size of the computational problem we consider a cracked plate symmetrically loaded by tensile step pulses on the edges $x_2 = \pm H$. Since this problem possesses reflective symmetry with respect to the crack plane, only the upper half plane needs to be considered. The solutions (5.2) and (5.3) are still applicable to this problem and by superposition, the magnitude of the dynamic K is twice that for the single step pulse.

For this analysis the upper half plane is modelled by 4-noded isoparametric elements. This choice is guided by our numerical experimentations on dynamically propagating cracks which indicate that the node release technique is better suited with 4-noded elements. The upper half plane model has 1290 elements (15 by 86 arrangement) and 1392 nodes. All elements are square shaped and identically sized. Homogeneous displacement boundary conditions are prescribed on the symmetry plane $x_2 = 0,\ x_1 \geq 0$.

To compare the convergence characteristics for the half plane model with the full plane model of the preceding section, calculations similar to those reported in Section 5.2 were carried out for a stationary crack. The convergence behavior with respect to element size and time steps for the present model is similar to that described in Section 5.2.

Nevertheless the results from the full plane model are in better agreement with the exact solution when equivalent meshes and identical time steps are employed. The solution obtained from the half plane refined mesh model (as described in the preceding paragraph) with uniform time steps of $0.25L/c_1$ and the exact solution are shown in Fig. 3. For $t > 0.2H/c_1$; the dynamic J agreed very well with the exact solution.

Before describing the main results, the algorithm for extending the crack is briefly discussed. Consider a crack propagating at constant speed v. The time required for the crack advance through one element length is L/v and denoted by t_e. At the start of the crack advance, the zero displacement boundary condition at the crack tip is replaced by an equivalent concentrated force, F_o, required to maintain the prescribed displacement. The nodal force is relaxed to zero in following increments according to a so-called "constant energy release rate" [29,30] relation,

$$F(t) = F_o\sqrt{1 - t/t_e}, \tag{5.5}$$

where t is the time elapsed from the start of the node release and $F(t)$ is the nodal force at time t. The merits of this nodal force relaxation scheme and other node relaxation schemes are discussed in [30]. We experimented with several nodal force relaxation schemes on several benchmark problems and reached the conclusion that the scheme according to (5.5) gave dynamic J results which are consistent with analytical solutions.

The problem to be analyzed is as follows. At time $t = -H/c_1$ a step pulse is applied at $x_2 = H$, and the longitudinal wave reaches the crack plane at $t = 0$. The crack tip is stationary for $t < \tau$, and it propagates at a constant speed v for $t \geq \tau$. Within the interval $-H/c_1 < t < \tau$, the calculations employ uniform time steps of $\Delta t = 0.25L/c_1$, where L/c_1 is the time required for the longitudinal wave to travel through an element. For $t \geq \tau$ the time steps are governed by the speed of the propagating crack v and the element size L. The crack advances across an element boundary (on $x_2 = 0$) in 20 increments, and this corresponds to uniform time steps of $\Delta t = L/20v$. The analyses were carried out for crack tip speeds of $0.4c_2$ and $0.6c_2$ respectively, where c_2 is the shear wave speed.

In the first analysis, the crack is stationary between $-H/c_1 < t < 0.467H/c_1$ and then prescribed to propagate at a uniform speed of $0.4c_2$. For the model employed, the time span where the numerical solution pertains to the semi-infinite crack in an infinite plane

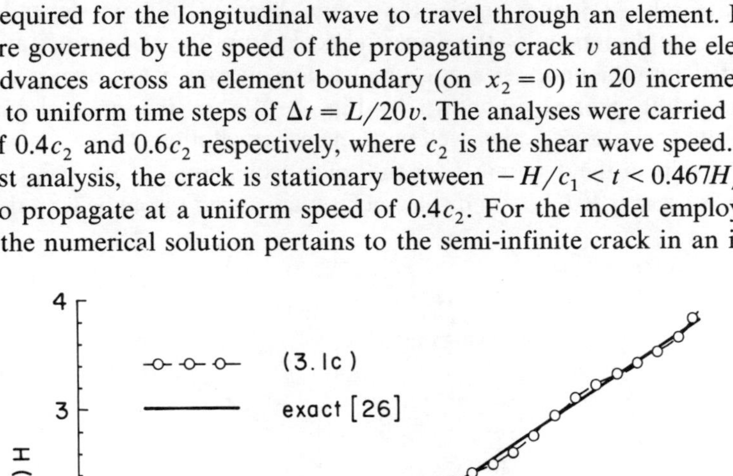

Figure 3. Comparison of normalized J as determined from the half plane model with the exact solution.

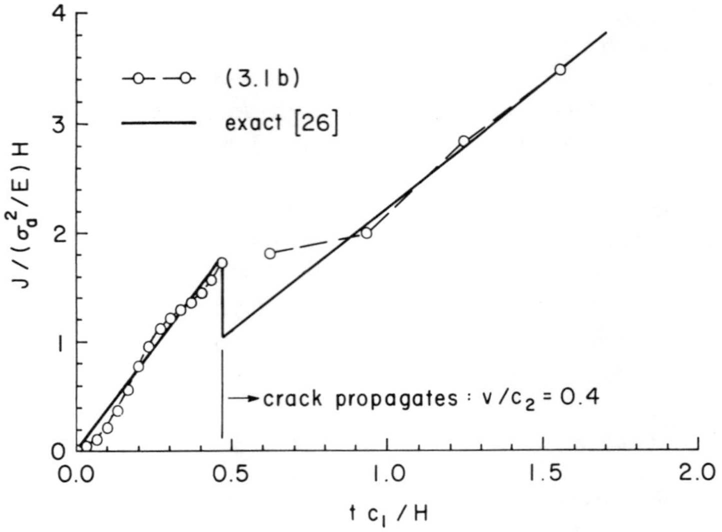

Figure 4. Comparison of normalized J with the exact solution. Crack propagates at speed $v/c_2 = 0.4$ at $t > 0.467H/c_1$.

problem is $-H/c_1 < t < 2H/c_1$. During this interval the crack tip advanced through four nodes. The dynamic J is evaluated at every time step. For $t < 1.667H/c_2$, the dynamic J values evaluated on various domains using (3.1b) are within several percent of its mean value. Thus the numerical results are consistent with the domain independence of the dynamic J integral. At longer times, the values become domain dependent and we attribute this to the dispersion of the waves reflected from the boundaries and numerical errors associated with the node release technique. The average of the dynamic J associated with each unloading process is shown in Fig. 4. Within the interval indicated, the numerical results agreed reasonably well with the exact solution.

Numerical results of the dynamic J for the crack propagating at $v = 0.6c_2$ are shown in Fig. 5. Here τ is set at $0.467H/c_1$ and the crack tip advanced through six nodes within the

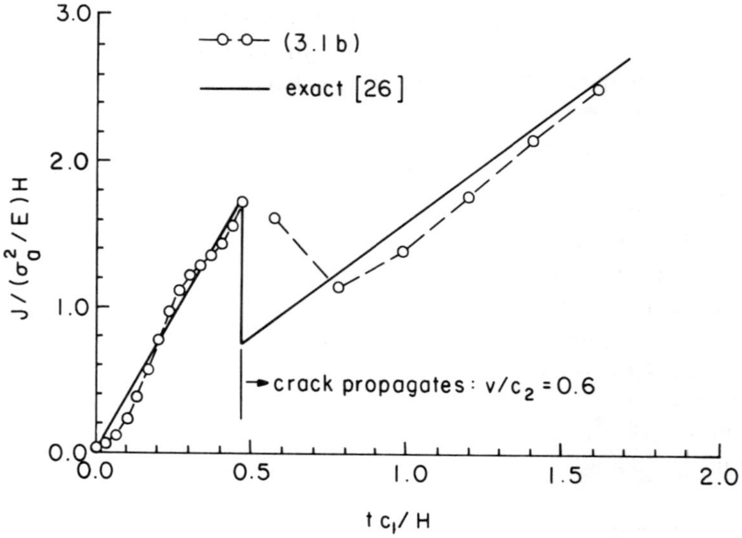

Figure 5. Comparison of normalized J with the exact solution. Crack propagates at speed $v/c_2 = 0.6$ at $t > 0.467H/c_1$.

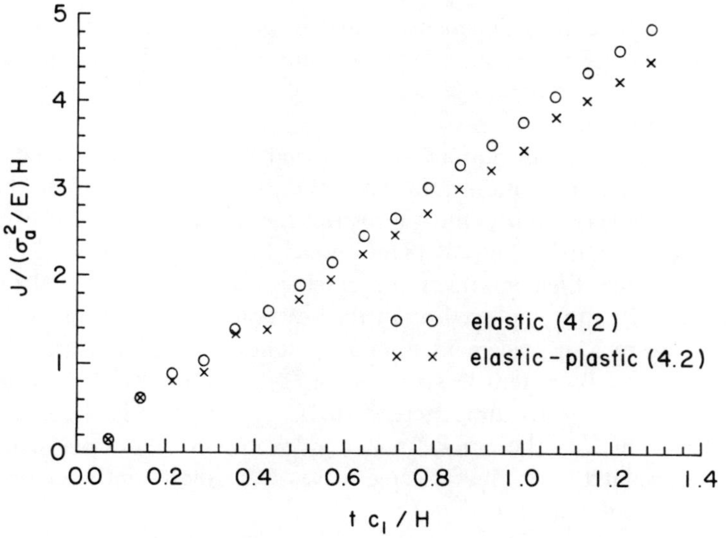

Figure 6. Behavior of dynamic J with respect to time as determined from elastic and elastic-plastic analyses.

period of interest. Again, the agreement between numerical results and the exact solution is reasonably good. It should be noted that the error expressed in terms of the dynamic stress intensity factor is only half of that for the dynamic J.

5.4. Elastic-plastic analysis of stationary crack subject to symmetrically applied step pulses

An elastic-plastic dynamic analysis is carried out using the half configuration discussed in Section 5.3 based on the following uniaxial tensile stress-strain relation,

$$\begin{aligned} \epsilon/\epsilon_o &= \sigma/\sigma_o, \qquad \epsilon < \epsilon_o \\ \epsilon/\epsilon_o &= (\sigma/\sigma_o)^n, \qquad \epsilon \geq \epsilon_o. \end{aligned} \tag{5.6}$$

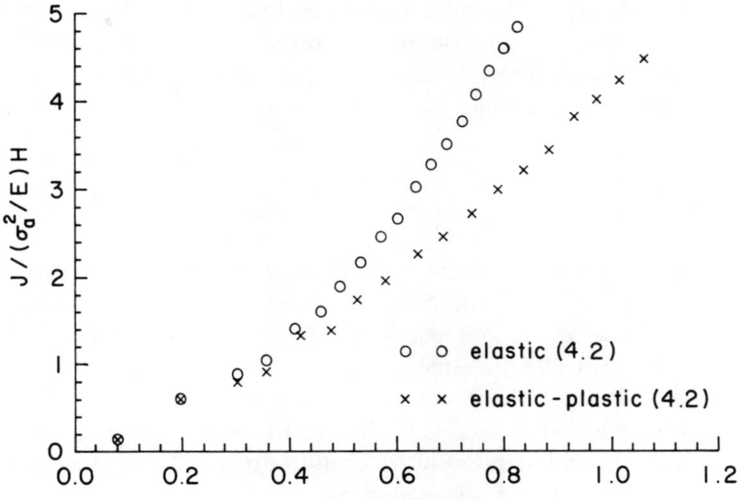

Figure 7. Behavior of dynamic J with respect to crack tip opening displacement as determined from elastic and elastic-plastic analyses.

Here n is the strain-hardening exponent, ϵ_o and σ_o are reference strain and stress related by $\sigma_o = E\epsilon_o$. In this particular study n has a value of 10 which corresponds to a lightly hardening material. The relation (5.6) is generalized to multiaxial stress states according to a J_2 flow theory of plasticity.

Elastic-plastic dynamic calculations were carried out for a mesh comprised of only 4-noded elements and another mesh comprised of only 8-noded elements. The results from both calculations are very similar and we restrict the discussion to results obtained from the 8-noded element model. The half plane model has 456 elements (12 by 38 arrangement) and 1469 nodes. Elements near the crack tip are square shaped and identically sized. Larger elements are employed at further distances from the crack tip. Symmetry conditions, that is, zero displacement boundary conditions are prescribed on the crack plane $x_2 = 0$ and $x_1 \geqslant 0$. A tensile step load σ_a of magnitude $0.6\sigma_o$ is applied on the boundary at $x_2 = H$. Uniform time steps of $0.5L/c_1$ where L is the size of the square elements in the vicinity of the crack tip is employed throughout the calculations. The calculation is terminated when the reflected waves from the boundaries reach the plastic zone emanating from the crack tip.

The dynamic J is computed according to (4.2). In this case the integration of the stress work density U follows the actual strain trajectory of a material point. (For a stationary crack, (3.1c) is identical to (4.2) if U in (3.1c) is taken to be the stress work density.) Again we observed that values evaluated on differently sized domains are within one percent of the main value throughout the entire analysis. This lends confidence to this particular elastic-plastic dynamic solution. The rise of dynamic J with respect to time is shown in Fig. 6. To gauge the effect of plasticity, we have included in the figure the elastodynamic J obtained from the same mesh using the identical time steps. The behavior of the dynamic J with respect to the opening displacement of the node at a distance L behind the crack tip as determined from the elastic and elastic-plastic analyses are shown in Fig. 7. The trend of these solutions are in accord with quasi-static results. An analytical solution to this problem is not available and more definite statements concerning the accuracy of the elastic-plastic numerical solution cannot be made. Nevertheless, results from another study [25] do provide information which is pertinent to the present discussion. In that study, a deeply notched round bar was subjected to a tensile pulse with a wave length which is long compared to the bar diameter. The dynamic J obtained from (4.2) agreed well with the values determined from a deep crack formula based on the remote load and displacement [31]. These earlier comparison taken together with the present results indicate that finite domain integrals can be employed to extract accurate values of the dynamic J from finite element solutions.

Acknowledgements

We gratefully acknowledge the research support of the Naval Sea Systems Command (SEA 05R15) through NBS Grant NB82RA20004, the National Science Foundation through Grant MEA82-10931, and the NSF Materials Research Laboratory at Brown University through Grant DMR83-16893.

The calculations described here were performed in the VAX-11/780 Computational Mechanics Facility at Brown University. This facility was made possible by grants from the National Science Foundation (Solid Mechanics Program), the General Electric Foundation, and the Digital Equipment Corporation.

References

[1] C. Atkinson and J.D. Eshelby, *International Journal of Fracture Mechanics* 4 (1968) 3–8.
[2] B.V. Kostrov and L.V. Nikitin, *Archiwum Mechaniki Stosowanej* 22 (1970) 749–775.
[3] L.B. Freund, *Journal of Elasticity* 2 (1972) 341–349.
[4] L.E. Malvern, *Introduction to the Mechanics of a Continuous Medium*, Prentice-Hall (1969) 210–211.
[5] J.D. Eshelby, in *Inelastic Behavior of Solids*, edited by M.F. Kanninen et al. McGraw-Hill (1970) 77–115.
[6] J.R. Willis, *The Mechanics and Physics of Fracture*, The Metal Society (1975) 57–67.
[7] K. Kishimoto, S. Aoki and M. Sakata, *Engineering Fracture Mechanics* 13 (1980) 387–394.
[8] S. Aoki, K. Kishimoto and M. Sakata, *Engineering Fracture Mechanics* 20 (1984) 827–836.
[9] S.N. Atluri, *Engineering Fracture Mechanics* 16 (1982) 341–364.
[10] T. Nishioka and S.N. Atluri, *Engineering Fracture Mechanics* 18 (1983) 1–22.
[11] T. Nishioka and S.N. Atluri, *Engineering Fracture Mechanics* 18 (1983) 23–33.
[12] T. Nishioka and S.N. Atluri, *AIAA Journal* 22 (1984) 409–414.
[13] L.B. Freund, in *Mechanics Today, Vol. III*, edited by S. Nemat-Nasser, Pergamon (1976) 55–91.
[14] J.W. Hutchinson, *Journal of the Mechanics and Physics of Solids*[15 16 (1968) 13–31.
[15] J.R. Rice and G.F. Rosengren, *Journal of the Mechanics and Physics of Solids* 16 (1968) 1–12.
[16] J.D. Achenbach and N. Nishimura, "Effect of inertia on finite near tip deformation for fast mode III crack growth", *Journal of Applied Mechanics* (to appear).
[17] J.K. Knowles and E. Sternberg, *Journal of Elasticity* 10 (1980) 81–110.
[18] J.R. Rice, *Journal of Applied Mechanics* 35 (1968) 379–386.
[19] J.R. Rice, in *Proceedings of the 8th US National Congress of applied Mechanics*, edited by R.E. Kelley, Western Periodicals, No. Hollywood, CA (1979) 191–216.
[20] J.W. Hutchinson, *Journal of Applied Mechanics* 50 (1983) 1042–1051.
[21] L.B. Freund and J.W. Hutchinson, "High strain-rate crack growth in rate dependent plastic solids", *Journal of the Mechanics and Physics of Solids* (to appear).
[22] ABAQUS, Hibbitt, Karlsson and Sorenson, Inc., Providence, Rhode Island.
[23] H.M. Miller, T.J.R. Hughes and R.L. Taylor, *Earthquake Engineering and Structural Dynamics* 5 (1972) 129–140.
[24] C.F. Shih and A. Needleman, *Journal of Applied Mechanics* 51 (1984) 48–56.
[25] T. Nakamura, C.F. Shih and L.B. Freund, "Elastic plastic analysis of a dynamically loaded circumferentially notched round bar", *Engineering Fracture Mechanics* (to appear).
[26] L.B. Freund, *Journal of the Mechanics and Physics of Solids* 21 (1973) 47–61.
[27] L.B. Freund, *Journal of the Mechanics and Physics of Solids* 20 (1972) 129–140.
[28] R.S. Barsoum, *International Journal on Numerical Methods in Engineering* 11 (1977) 85–98.
[29] G. Rydholm, B. Fredriksson and F. Nilsson, in *Numerical Methods in Fracture Mechanics*, edited by A.R. Luxmoore and D.R.J. Owens, University of Swansea, Swansea, Wales (1978) 660–672.
[30] J.F. Malluck and W.W. King, in *ASTM STP* 711, American Society for Testing and Materials (1980) 38–53.
[31] J.R. Rice, P.C. Paris and J.G. Merkle, in *ASTM STP* 536, American Society for Testing and Materials (1973) 231–245.

Résumé

On passe en revue les développements dans l'utilisation de l'intégrale du flux d'énergie à l'extrémité d'une fissure dans le calcul de la mécanique de la rupture dynamique et ce, au cours des 5 dernières années. On en tire une expression du flux d'énergie à l'extrémité d'une fissure en fonction du champ mécanique au voisinage de cette extrémité, expression applicable aux réactions générales d'un matériau. On démontre ensuite que certaines intégrales d'énergie intéressantes peuvent être extraites des résultats généraux en mettant en exergue les caractérisations adéquates de la réponse du matériau. Diverses représentations alternatives du flux d'énergie sous forme d'intégrale couvrant certaines régions finies autour de l'extrémité de la fissure sont présentées et comparées, en vue d'une insertion dans des études de simulation per éléments finis.

International Journal of Fracture 27 (1985) 245–261.

Numerical studies in dynamic fracture mechanics

SATYA N. ATLURI * and T. NISHIOKA **

* *Center for the Advancement of Computational Mechanics, Georgia Institute of Technology, Atlanta, GA 30332, USA*

** *Dept. of Ocean Mechanical Engineering, Kobe University of Mercantile Marine, 5-1-1 Fukaeminamicho, Higashinada-Ku Kobe, 658, Japan*

(Received September 1, 1984; in revised form November 8, 1984)

Abstract

This paper provides a summary of recent studies concerning numerical modeling of dynamic crack-propagation. Both "stationary mesh" as well as "moving mesh" finite-element procedures are examined. Simple procedures, using a moving mesh of conventional isoparametric elements in conjunction with certain path-independent integrals for the evaluation of stress-intensity factors for a dynamically propagating crack are presented.

1. Introduction

Dynamic fracture mechanics concerns (i) the onset of crack growth under dynamic (impact) loadings and (ii) dynamic (unstable) crack propagation in stressed solids. The conventional linear elasto-static or quasi-static fracture mechanics applies only up to the end of stable growth under quasistatic loadings and assumes that the onset of unstable crack propagation renders the structure useless. Yet, the prevention of crack-growth initiation itself may be too conservative or too costly an objective for some structural designs. On the other hand, the catastrophic failures caused by unstable crack growth are also obviously intolerable. In these cases, the assurance of crack arrest, as a second line of defense, is essential. This concept has been investigated for design methodologies assuring the integrity of nuclear pressure vessels under thermal shock conditions, LNG ship hulls, and gas transmission pipelines. These studies were reviewed in [1].

The fundamental framework of the subject of dynamic fracture mechanics relies primarily on solutions for dynamic behavior of solids containing cracks. These solutions are characterized by (i) stress wave interactions, (ii) the need to account for kinetic energy in the global energy balance of the fracturing body, and (iii) inertia effects of the material. In the past two decades or so, a number of analytical solutions, which provide a useful understanding of dynamic crack behavior, have been obtained. These analytical solutions are, however, limited to cases of simple loadings and unbounded plane bodies. Moreover, the stress wave interactions, which play an important role in dynamic fracture mechanics, usually render the analytical solutions intractable. Therefore, the use of numerical methods is often indispensible for the analyses of cracks in finite solids.

However, the ability to perform an elastodynamic analysis alone is not enough to assess the situation of dynamic fracture in a solid. Values of the material's resistance to crack propagation, i.e., the dynamic fracture toughness, must also be available. Unfortunately, for materials such as steel, a direct experimental measurement of dynamic fracture toughness, as a function of crack velocity, is often very difficult. To overcome these difficulties, hybrid experimental-numerical methods of analysis of laboratory tests are often preferred [2,3]. This is another reason why the advancement of dynamic fracture mechanics relies heavily on advances in computational methods for the analysis of propagating cracks.

Numerical methods in dynamic fracture mechanics were critically appraised in 1978 [4]. At that time, a comparison of the finite element and finite difference methods led to the following conclusion. The finite element method was more suitable, for the analysis of stationary cracks under dynamic loadings, due to the fact that the relevant singularities can be modeled in the crack-tip elements. On the other hand, the finite element method was thought to be unsuited for the analysis of dynamic crack propagation, due to the numerical difficulties involved in advancing the crack in a discrete manner.

However, the state of the art of finite element methods in dynamic fracture mechanics has greatly advanced in the intervening years. In large measure, this is the result of the development of finite-element procedures to model propagating singularities and of path-independent integrals which characterize the strength of the fields near a propagating crack-tip. This article is devoted primarily to a review of these developments.

2. Basic concepts of dynamic fracture mechanics

2.1. Crack-tip behavior

A logical classification of dynamic fracture problems may be stated as follows: (i) stationary cracks under dynamic loading, (ii) dynamically propagating cracks under quasi-static loading, and (iii) dynamically propagating cracks under dynamic loading. The elastodynamic fields near the tip of a stationary crack have the same form as those in elasto-statics [5,6]:

$$\sigma_{ij}^0 = \sum_M \frac{K_M}{\sqrt{2\pi r}} f_{ij}^M(\theta) \qquad (M = \text{I, II, III}) \tag{1}$$

$$u_i^0 = \sum_M \frac{K_M}{2\mu} \sqrt{\frac{r}{2\pi}}\, g_i^M(\theta) \qquad (M = \text{I, II, III}), \tag{2}$$

where $M = \text{I, II, III}$ denotes the "mode" of fracture, μ is the shear modulus, and (r, θ) are polar coordinates centered at the crack-tip. The superscript "0" in $(\ )^0$ denotes the tensor components $(\ )$ with respect to the local coordinates x_k^0 attached to the crack-tip (see Fig. 1). The dynamic stress intensity factors K_M for stationary cracks depend on time t, crack length a, loading rate $\dot{\sigma}$, as well as the boundary conditions.

As a typical example, we consider a semi-infinite stationary crack subjected to a (normal) step wave of stress σ_M ($\sigma_{\text{I}} = \sigma_{22}^0$, $\sigma_{\text{II}} = \sigma_{12}^0$, $\sigma_{\text{III}} = \sigma_{23}^0$) [5,7,8]. The stress intensity factors vary with time as:

$$K_M = D_M(\rho, \mu, \nu)\sigma_M\sqrt{t} \qquad (M = \text{I, II, III}), \tag{3}$$

where D_M depend only on material properties, with ρ being the mass density and ν Poisson's ratio. The time t is taken to be zero at the instant when the stress waves reach the crack-tip.

For dynamically propagating cracks under dynamic or quasi-static external loading, the crack velocity $C(\equiv \dot{a})$ establishes the stress and displacement fields as follows [9–12]:

$$\sigma_{ij}^0 = \sum_M \frac{K_M(C)}{\sqrt{2\pi r}} f_{ij}^M(\theta, C) \tag{4}$$

$$u_i^0 = \sum_M \frac{K_M(C)}{2\mu} \sqrt{\frac{r}{2\pi}}\, g_i^M(\theta, C). \tag{5}$$

The general solutions for the stress and displacement fields near the dynamically propagating crack-tip were obtained in [12] in a unified fashion for all three fracture modes. The stress intensity factors K_M in (4, 5) may be expressed as products of velocity factors $k_M(C)$ and "static" factors K_M^*, as noted by several researchers [13,14,15]:

$$K_M(t, C) = k_M(C) K_M^*(t) \qquad (M = \mathrm{I, II, III}), \tag{6}$$

The "static" factors K_M^* depend on the current length of the crack, the applied load, the history of crack extension, but not on the instantaneous crack velocity. As discussed in [14] and [15], K_M^* are, in general, not equal to the static stress intensity factors, $K_{M_{\mathrm{sta}}}$, for a stationary crack of the same length as the moving crack.

For a crack propagating in a self-similar fashion at an angle θ_0 measured from the global X_1 axis as shown in Fig. 1, the energy release rate G can be expressed as [16]:

$$G = (C_k/C) G_k = G_1 \cos\theta_0 + G_2 \sin\theta_0 \tag{7}$$

and

$$G_k = \lim_{\epsilon \to 0} \int_{\Gamma_\epsilon} [(W + K) n_k - T_i u_{i,k}] \, \mathrm{d}S, \tag{8}$$

where C_k are the components of crack velocity in the X_k system, and G_k is the energy-release rate due to a unit crack translation along the global coordinate X_k; Γ_ϵ is a contour arbitrarily close to the propagating crack-tip; W and K are the strain and kinetic energy densities, respectively; n_k the outward normal direction cosines; T_i the traction; and $(\ )_{,k}$ denotes $\partial(\ )/\partial X_k$.

We now consider the change of components of G under a coordinate transformation. Suppose that the angle between the X_1 and x_1^0 axes is θ_0 (see Fig. 1). The local Cartesian coordinates x_i^0 can be expressed by

$$x_i^0 = \alpha_{ij}(\theta_0) X_j; \quad \alpha_{ij}(\theta_0) = \begin{bmatrix} \cos\theta_0 & \sin\theta_0 \\ -\sin\theta_0 & \cos\theta_0 \end{bmatrix}. \tag{9}$$

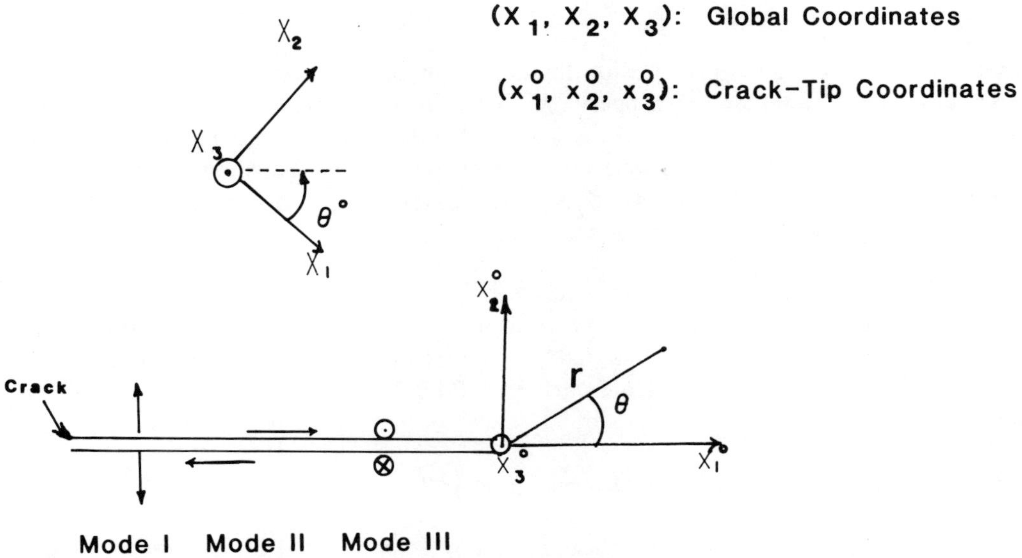

Figure 1. Nomenclature for mixed mode loading.

The same transformation holds for the components of energy release rate G_k. Thus, we have

$$G_k^0 = \alpha_{kl} G_l \quad \text{or} \quad G_k = \alpha_{lk} G_l^0. \tag{10}$$

Substituting the singular stress and corresponding displacement fields given by (4) and (5) into (8) with θ_0 being zero, the relations between the energy release rate and the instantaneous stress intensity factors can be expressed as [12]:

$$G_1^0 = \frac{1}{2\mu}\left\{ K_{\mathrm{I}}^2 A_{\mathrm{I}}(C) + K_{\mathrm{II}}^2 A_{\mathrm{II}}(C) + K_{\mathrm{III}}^2 A_{\mathrm{III}}(C) \right\} \tag{11a}$$

$$G_2^0 = -\frac{K_{\mathrm{I}} K_{\mathrm{II}}}{\mu} A_{\mathrm{IV}}(C). \tag{11b}$$

The crack velocity functions, $A_{\mathrm{I}}(C) \cdots A_{\mathrm{IV}}(C)$ are given in [12].

2.2. Dynamic fracture criteria

The criteria governing the motion of the crack (growth initiation, propagation, and arrest) can be expressed, when linear elastic conditions prevail, in terms of the stress intensity factor K and the experimentally determined critical values that are taken to be material specific.

In the conventional, static LEFM, the condition for the onset of crack growth is often expressed as

$$K = K_c, \tag{12}$$

where K_c is called the static fracture toughness.

In elastodynamic fracture, one has to consider two counterparts of fracture toughness of materials [1]. First, for the onset of growth of a rapidly loaded stationary crack, we have

$$K = K_d(\dot{\sigma}) \qquad (\text{or} \quad K = K_d(\dot{K})), \tag{13}$$

where K_d is the dynamic "initiation" toughness, K_d may also depend on the prevailing temperature. For the limiting case of $\dot{\sigma} = 0$, K_d is identical with K_c. Second, for dynamically propagating cracks under quasi-static as well as dynamic loading conditions, it has been suggested that

$$K = K_{\mathrm{D}}(\dot{a}), \tag{14}$$

where K_{D} is the dynamic propagation toughness for propagating cracks. There is some controversy as to whether the property $K_{\mathrm{D}}(\dot{a})$ is truly geometry-independent. On the other hand, for viscoelastic materials K_{D} may also depend on the crack acceleration $\ddot{a}$ and/or other higher-order time derivatives of crack length. According to the criterion given by (14), crack arrest occurs when the stress intensity factor becomes smaller than or equal to a critical value. This can be expressed as

$$K \leqslant K_{\mathrm{D}}(0) \equiv K_{\mathrm{Ia}}^{\mathrm{dyn}}, \tag{15}$$

where $K_{\mathrm{Ia}}^{\mathrm{dyn}}$ denotes the dynamic crack arrest toughness. In the above, the superscript "dyn" is used to distinguish a material property from the so-called arrest toughness $K_{\mathrm{Ia}}(\equiv K^{\mathrm{sta}})$, which is utilized in the current ASME Boiler and Pressure Vessel Code, Section XI.

3. Finite element modeling of dynamic crack propagation

In the finite element method, the solid continuum is represented by a mesh of a finite number of elements. To simulate the crack propagation, two different concepts of computational modeling may be considered, namely, stationary (fixed) mesh procedures and moving (distorting) mesh procedures. For each procedure, several alternate schemes have been presented in literature.

3.1. Stationary mesh procedure

We review first the feature of the stationary mesh procedure. In early applications of the finite-element method to dynamic crack propagation [17,18], the crack-tip motion was modeled by discontinuous jumps. This was done by changing the location of the crack-tip from one node to the next ("node-shifting" procedure) along the crack axis, in time Δt employed in the time integration scheme. To obtain accurate finite element solutions for wave propagation problems, one must use a small time increment Δt – usually the time for dilatational stress waves to travel the distance between the two closest finite element nodes. Since the velocity of crack propagation is usually significantly lower than the wave velocity, the crack-tip will move during this time increment Δt to a position somewhere in between the two successive nodal points. Also the sudden increase of crack length, and the release of constraint on the displacement, induce spurious high-frequency oscillations in the finite element solutions. To overcome these difficulties, several attempts have been made at releasing the nodes gradually, over a period of time. These are discussed below.

Keegstra et al. [19,20] have proposed a model, as shown in Fig. 2a, which allows for the presence of non-zero "holding-back forces" at more than one node behind the advancing crack-tip. Yagawa et al. [21] us a Lagrange multiplier approach to enforce the displacement boundary condition on the "uncracked" part of the surface of the crack-tip element [marked as segment (CD) in Fig. 2b] in situations when, in time Δt, the crack-tip actually propagates to a location in between two adjacent nodes. Note that the traction T_2, as shown in Fig. 2b, is used as a Lagrange multiplier on segment CD. Other "gradual" nodal release mechanisms were presented in the literature, with different rates of release of nodal

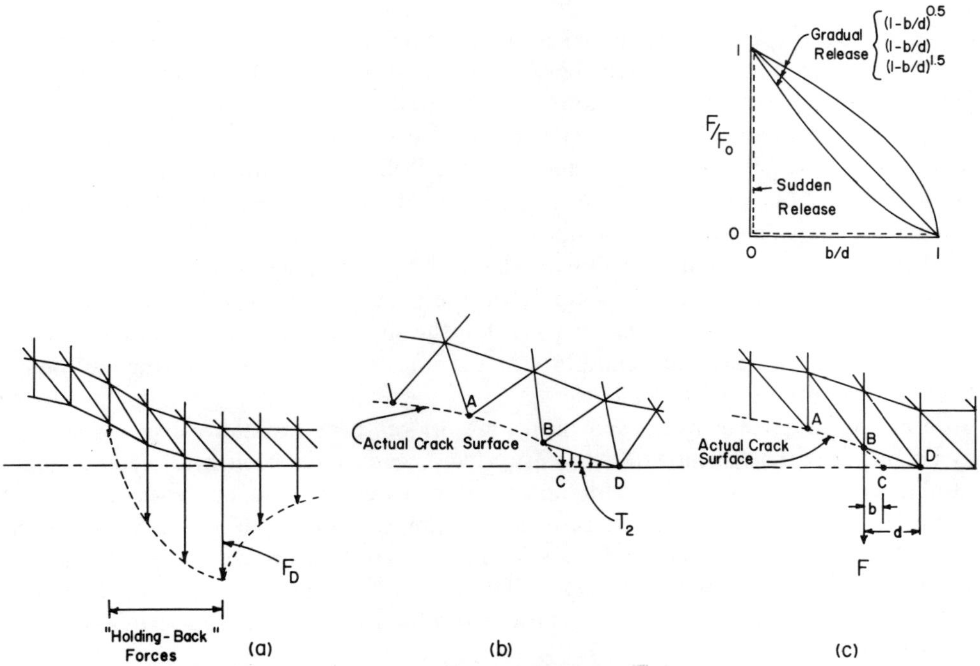

Figure 2. "Gradual-nodal-release" techniques.

forces. Suppose that the actual crack-tip is located at point C in between the finite element nodes B and D as shown in Fig. 2c, and that "b" and "d" are the lengths of segments (BC) and (BD), respectively. The "holding-back" force F at node B is gradually reduced to zero as the crack-tip reaches the node D. The various schemes for accomplishing this are summarized below:

Malluck and King [22] $F/F_0 = (1 - b/d)^{1/2}$ (16a)

Rydholm, Fredriksson and Nilsson [23] $F/F_0 = (1 - b/d)^{3/2}$ (16b)

Kobayashi, Mall, Urabe and Emery [24] $F/F_0 = (1 - b/d)$, (16c)

where F_0 is the original reaction force when the crack-tip is located at the node B. It is important to recognize that the above procedure of moving the crack-tip from B to D involves several time-steps Δt. In the use of eight-noded isoparametric elements, it was demonstrated in [25] that both the corner and midside nodes on the crack element should be released simultaneously. Studies using different nodal release mechanisms, as in (16a–c), were discussed in [26,27]. From the authors' experience, the use of "linear relaxation" (as in (16c), while employing eight-noded elements, yields smoother results than the other "nonlinear" relaxation techniques.

3.2. Moving mesh procedure

Next we present a synopsis of the moving (shifting) mesh procedures in which the mesh moves with the crack-tip.

Bazant, Glazik, and Achenbach [28] developed a non-singular finite element method for moving the entire mesh with the crack-tip. However, this procedure has two obvious limitations: (i) it is restricted to infinite bodies whose surfaces and/or bimaterial interfaces are parallel to the direction of crack propagation; and (ii) more importantly, it cannot be applied to bodies having finite dimensions in the direction of crack propagation. An early attempt toward a singular element for propagating cracks was made by Aberson, Anderson and King [29,30], who used the eigenfunctions appropriate for a stationary crack [31] as basis functions in the propagating singular element. The crack-tip moves within the singular element, between the nodes A and B as shown in Fig. 3(a). When the crack-tip reaches the node B, the mesh pattern is changed suddenly, as illustrated in the figure. However, it has been reported [32] that this procedure did not yield meaningful results. Another attempt at using Williams' eigenfunctions [31] was made by Patterson and Oldale [33,34]. The singular element [see Fig. 3(c)] has 13 nodes and is topologically equivalent to two assembled eight-noded isoparametric elements. The location of this singular element is suddenly changed by a distance equal to the size of the regular elements ahead of the crack-tip, when the crack-tip propagates a critical distance (about 80% of the length of the element) within the singular element. This model violates the displacement compatibility condition at the interfaces between the singular element and the surrounding regular elements.

In both the singular elements shown in Figs. 3(a) and (c), for stationary or quasi-statically propagating cracks, the stress intensity factor can be determined directly as one of the unknown coefficients of Williams' stationary-crack eigenfunctions. However, since the near-tip fields depend on crack velocity for dynamically propagating cracks, the coefficient of the singular eigenfunction for a stationary–crack does not correspond to the stress intensity factor for a propagating crack. Thus, care must be exercised in inferring the dynamic stress intensity factors. An empirical formula for this purpose was derived by the authors in [35].

Aoki, Kishimoto, and Sakata presented a singular element procedure [36,37] wherein

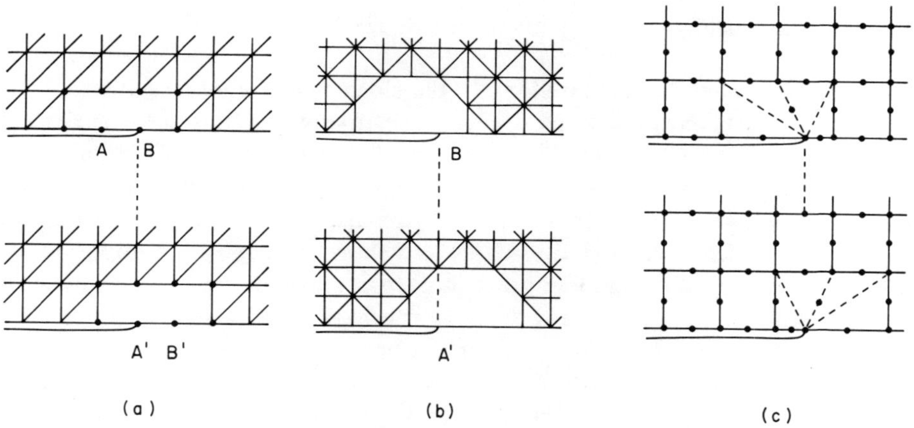

Figure 3. Discontinuously moving singular elements.

only the appropriate rigid motion and the eigenfunction corresponding to the singular stress field for a propagating crack were incorporated. The crack propagates in the singular element until it reaches the point B indicated in Fig. 3(b). Then the singular element suddenly changes its location as illustrated in the bottom half of Fig. 3(b). Since the size of the element in [36,37] is in general much larger than the region of validity of the singular solution alone, noticeable errors may occur in the determination of stress intensity factors. Omission of the constant stress field and higher order stress fields also limits the applicability of such an element in studying physical problems of interest such as crack-branching, etc.

In developing a special element, certain simple requirements have to be met in order to ensure the convergence of the solution. The displacement field in the element must: (i) include the appropriate rigid body motions, (ii) be able to represent a constant strain (stress) field, and (iii) be continuous at the interfaces between elements. In addition, to avoid kinematic modes in the special element, i.e. zero energy modes other than rigid modes, certain requirements on the number of parameters in the element basis functions have to be satisfied. Suppose a special element has N unknown parameters, i.e., N eigenfunctions including appropriate rigid body motions and M unconstrained nodal degrees of freedom. It can then be shown [35] that one has to satisfy an inequality, $N \geqslant M$. The developments in [29,30,33,34,36] and [37] violate some or all of these criteria.

A moving singular element procedure which satisfies all the above criteria for convergence and which employs eigenfunctions [12,40] for a dynamically propagating crack, has recently been extensively developed [38,39]. Thus, in the singular element, it is assumed

$$u_i^s(x_1^0, x_2^0, t) = U_{ij}(x_1^0, x_2^0, C)\alpha_j(t) \quad (i = 1, 2; \quad j = 0, 1, 2, \ldots, N-1), \tag{17}$$

where (x_1^0, x_2^0) is the moving coordinate ⋆ system shown in Fig. 1. The functions U_{ij} include the zero stresses and rigid body motions ($j = 0$), the singular stresses and corresponding displacements ($j = 1$), the constant stresses and linear displacements ($j = 2$), and the higher-order terms ($j \geqslant 3$). Thus, the coefficient α_1 is equivalent to the dynamic stress intensity factor. The displacement functions given by (17) satisfy the requirements (i) and (ii) mentioned above for the convergence of the solution. The compatibility of displacements, velocities, and accelerations at the boundary between the singular element

⋆ The coordinate system x_1^0, x_2^0 is assumed to be always centered at the crack-tip, and to move with the crack-tip velocity C.

and regular elements (Type B elements in Fig. 4) can be satisfied through a least-square approach [38,39].

In the "moving mesh" procedure [38,39], the singular element translates in each time step for which crack growth occurs as illustrated in Fig. 4. Thus the crack-tip always remains at the center of the singular element throughout the analysis. The regular elements (the type B elements in Fig. 4) surrounding the moving singular element are continuously distorted. To simulate a large amount of crack propagation, the mesh pattern around the moving element is periodically readjusted as also shown in Fig. 4. An important feature which distinguishes the moving-mesh procedures from the nodal release techniques is that the size of the increment of crack growth is not affected by nodal spacing but can be made as small as desired. Therefore, the displacement boundary condition ahead of the actual crack-tip can be satisfied exactly in the moving mesh procedures. Another attractive feature of the moving singular element procedure [38,39] is that the dynamic stress intensity factor at each time step can be calculated directly as the coefficient of one of the eigenfunctions, e.g. α_1 for a mode I case.

To simplify the moving singular element procedure, a study was undertaken in [35] and [41] to address the effect of alternatively modeling the moving region A in Fig. 4 with either the "quarter-point" singular elements (model A′) or the regular (non-singular) isoparametric elements (model A″). In the moving (non-singular) isoparametric element procedure, one has to calculate the stress intensity factors indirectly. A most accurate

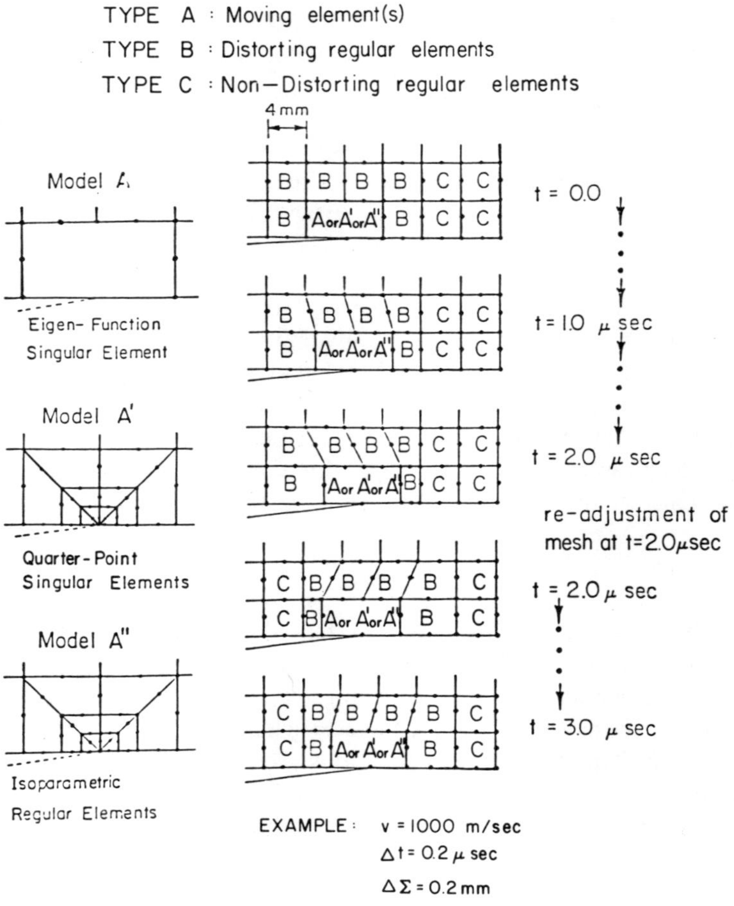

Figure 4. Moving element procedures.

estimate can be obtained by using certain path-independent integrals as discussed later in Section 5. While simplified procedures may lead to acceptable estimates of "overall parameters" like the stress-intensity factors, the use of more refined approaches, such as the moving singular element approach detailed above, is still an indispensible tool for purposes of basic research into fracture phenomena such as crack branching, etc.

4. Finite element simulation of dynamic fracture

Since a computational model can only respond in a manner as prescribed by input data, numerical simulations of dynamic fracture in materials require experimentally determined information. This approach may be labeled as the hybrid experimental-numerical method [2,3]. Computational simulations of dynamic crack propagation and crack arrest, for a specified initial flaw size, specimen geometry, and applied load, can be conducted in either of two different ways [4]. One of these is the so-called "generation phase simulation" in which the variation of stress intensity factor can be determined, using an experimentally measured crack-propagation history (a vs. t and/or C vs. t) as the input data into the computational model. From this calculation, one determines the dynamic fracture toughnesses through (13) and (14). Once the material properties K_d and K_D, as defined earlier, are determined numerically or experimentally, they may be used in the second type of computational simulation, a "prediction phase simulation". In this calculation, the crack propagation history (a vs. t) can be determined by specifying the initial conditions and material fracture toughness data as inputs to the computational model. The prediction phase simulation is also sometimes called "application", "propagation" or "inverse" simulation.

In the following, we consider the finite element simulation of dynamic crack propagation and arrest in a Wedge-loaded Rectangular Double Cantilever Beam (RDCB) specimen (see Fig. 5). Due to symmetry, only the upper half of the specimen is modeled by finite elements. The moving singular element is shown hatched at the beginning of crack propagation. In the experiment conducted by Kalthoff et al. [42], several test specimens were studied, wherein cracks were initiated from blunt notches, each with an apparent initiation stress intensity factor K_{IQ} larger than the fracture toughness K_{IC}. In their

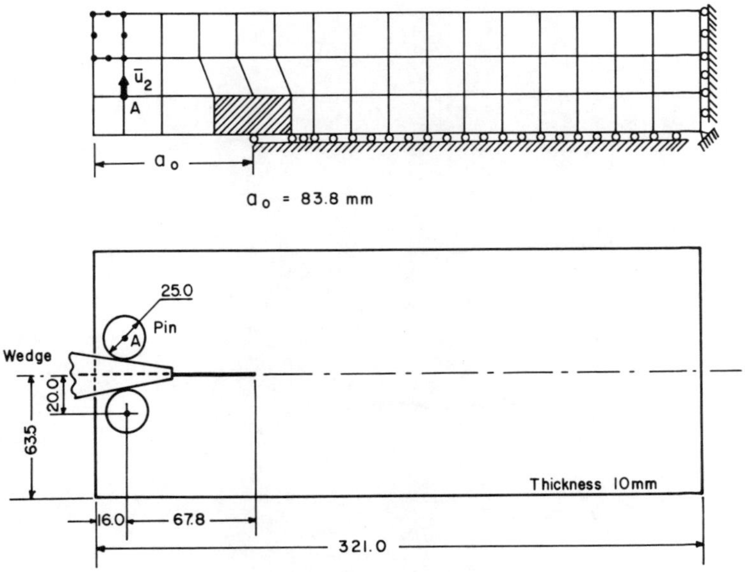

Figure 5. Finite element mesh at $t = 0$, of DCB specimen.

report [42] the RDCB specimens with K_{IQ} values of 2.32 and 1.33 Mpa · $m^{1/2}$ were identified, respectively, as specimens with numbers 4 and 17. For convenience, the same identifications are used here. The specimens were made of a photoelastic material with the properties of $E = 3.38$ Gpa, $\nu = 0.33$.

First we present the results of generation phase simulation for No. 4 Specimen[43]. The experimentally measured crack length and crack velocity history curves, i.e., $a = a(t)$ and $C = C(t)$ as given in Fig. 6, were used as the input data for the generation phase simulation. The variation of the computed K_I values agrees very well with the experimentally measured (using an optical method) data.

In the prediction phase simulation, the crack velocity in each time step will be determined from the given fracture toughness versus crack velocity relation, i.e., (14). Before proceeding with the prediction phase simulation from time t_1 to $t_2(= t_1 + \Delta t)$, one has to know the position of crack-tip at time t_2. To this end, the crack velocity at time $t_1 + \Delta t/2$ is predicted by using the following Taylor series expansion:

$$K_{Ip}\left(t_1 + \frac{\Delta t}{2}, C + \Delta C\right) = \sum_{n=0}^{\infty} \frac{1}{n!}\left\{\left(\frac{\Delta t}{2}\right)\frac{\partial}{\partial t} + \Delta C\frac{\partial}{\partial C}\right\}^n K_I(t_1, C_1), \tag{18}$$

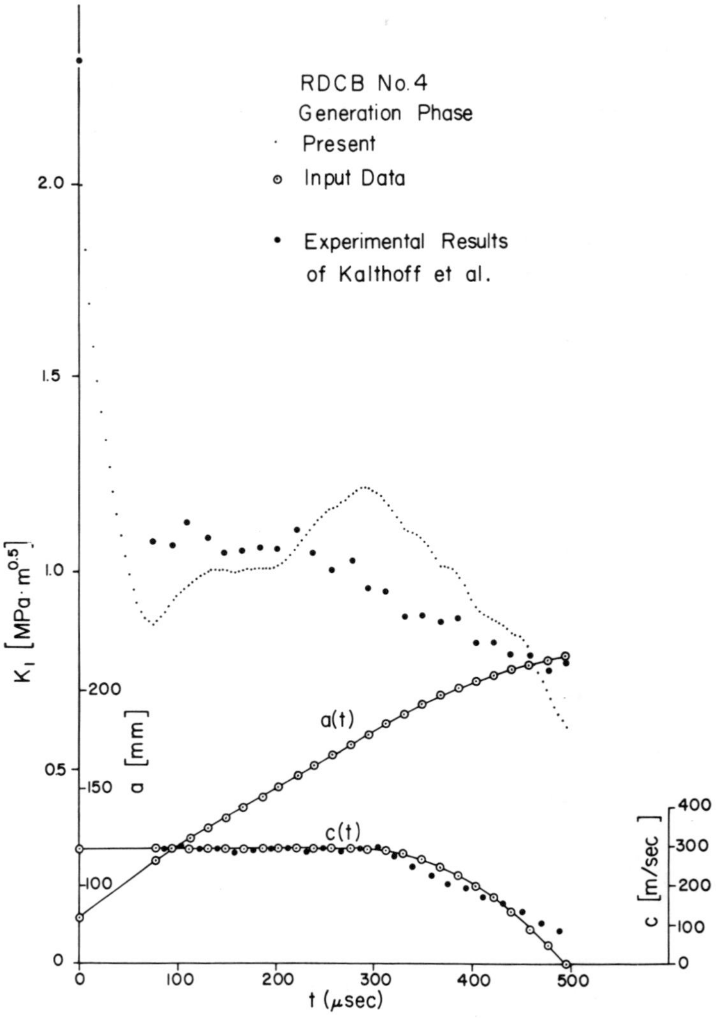

Figure 6. Results of generation phase simulation (No. 4).

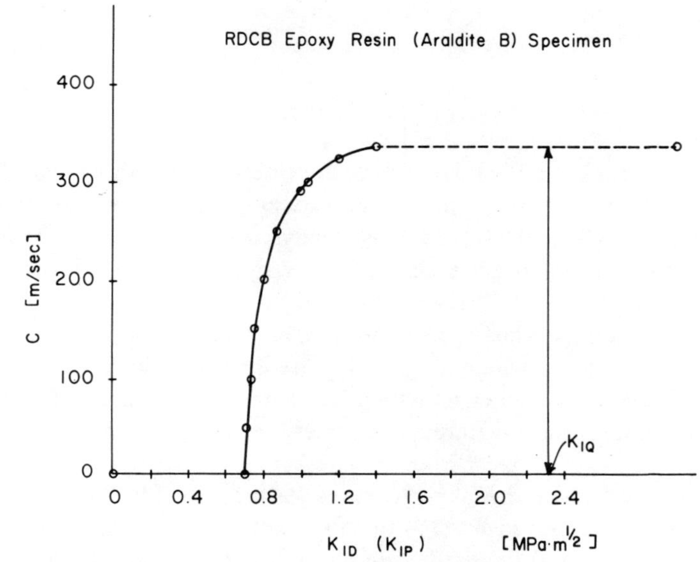

Figure 7. Fracture toughness versus crack velocity relation measured experimentally. [42].

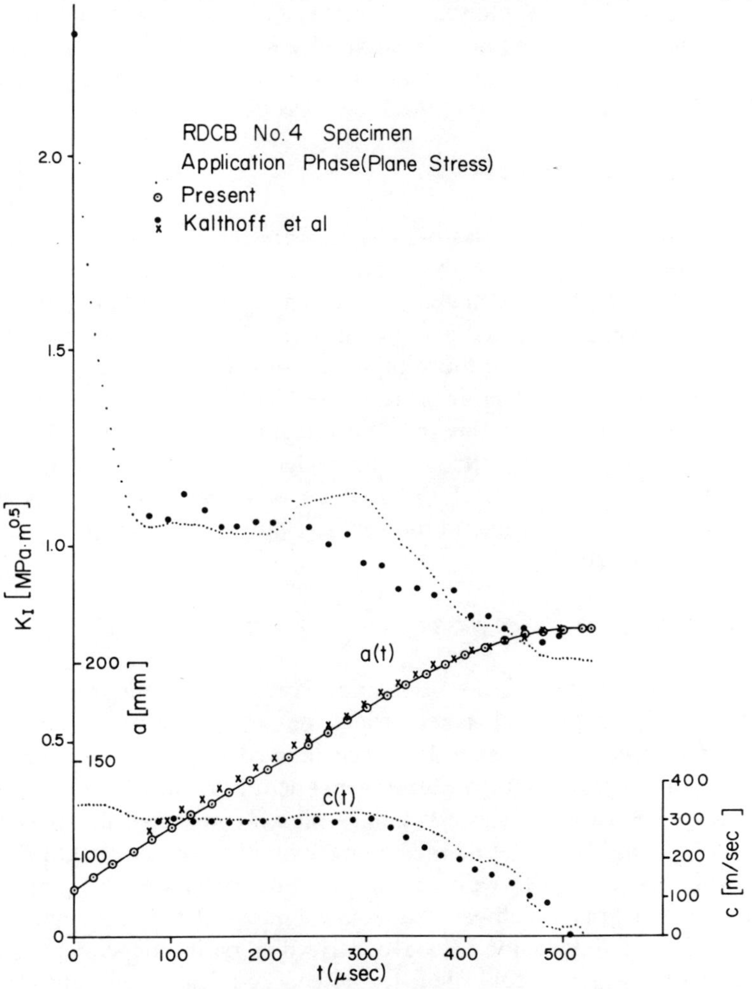

Figure 8. Results of prediction phase simulation (No. 4).

where K_{Ip} is the predicted K_I at $t_1 + \Delta t/2$ while $K_I(t_1, C_1)$ is known. One may now rewrite (18) as:

$$K_{Ip} = K_I(t_1, C_1) + \left(\frac{\Delta t}{2}\right)\dot{K}_I(t_1, C_1) + \frac{(\Delta t)^2}{8}\ddot{K}_I(t_1, C_1) + R, \tag{19}$$

where R is the residue; alternatively, it may be considered as a corrector of the prediction. The form of R depends on the type of variational principle used in the analysis, as discussed in detail in [43] and [44]. In most analyses based on the principle of virtual work, a zero corrector ($R = 0$) gives enough accuracy in the numerically calculated K_I values.

In the moving singular element procedure [38,39], one has $K_I(t_1) \equiv \alpha_1(t_1)$; $\dot{K}_I(t_1) \equiv \dot{\alpha}_1(t_1)$; and $\ddot{K}_I(t_1) \equiv \ddot{\alpha}_1(t_1)$ in which $\underset{\sim}{\alpha}$, $\underset{\sim}{\dot{\alpha}}$, and $\underset{\sim}{\ddot{\alpha}}$ are directly calculated as the parameters for the singular element. For other non-singular element procedures, $\dot{K}_I$ and $\ddot{K}_I$ can be determined using simple finite difference schemes, for example, $\dot{K}_I \simeq [K_I(t_1) - K_I(t_1 - \Delta t)]/\Delta t$, etc.

The experimentally measured K_{ID} vs. C curve for the RDCB specimen [42] is shown in Fig. 7. Using a K_{Ip} value given by (19) in this experimental data, the crack velocity at $(t_1 + \Delta t/2)$ was determined in the prediction phase simulation [43]. The simulation results for No. 4 specimen with $K_{IQ} = 2.32$ Mpa $\cdot$ m$^{1/2}$ are shown in Fig. 8. The variation of stress intensity factors and the computed crack propagation and crack velocity histories are compared with those determined experimentally by Kalthoff et al. [42]. In the prediction phase simulation, the crack was assumed to have arrested when $K_{Ip} \leqslant K_{Ia}^{dyn} \equiv K_{ID}(0)$, where K_{Ia}^{dyn} is the dynamic arrest fracture toughness of the material.

5. Use of path-independent integrals

It is well known that in mode I elasto-static fracture the J integral can be used for an efficient and simple computation of the relevant crack-tip parameters. This is so because, under appropriate assumptions of material homogeneity, etc., the strength of the crack tip fields (K_I, for instance) is governed also by an integral evaluated over a path that is far removed from the crack tip. Since the computed stress and displacement data in regions far removed from the crack tip is, in general, relatively insensitive to the details of the modeling of the crack tip region, this far-field integral can be evaluated with reasonable accuracy using relatively coarse finite element models of the structure. In this section, we discuss the computational details of the analysis of dynamic crack propagation using "path-independent" integrals which, however, also involve domain-integrals, due to the presence of material inertia.

5.1. Path independent-integrals for elastodynamically propagating cracks

Through an elaborate discussion of conservation laws, a path-independent integral J for an elastodynamically propagating crack, which has the meaning of rate of change of Lagrangean per unit crack extension, has been derived in [16]. Later, through a simple modification of an integral in [16], a path-independent integral J', which is equivalent to the rate of energy release G in elastodynamic crack propagation, has been derived [12]. The integrals in [12] and [16] were defined along spatially fixed paths. On the other hand, using paths which move rigidly with the propagating crack tip, Bui [45] has derived a path-independent integral which is also equivalent to the energy release rate. Also Kishimoto, Aoki, and Sakata [46,47] have derived a path-independent integral $\hat{J}$ for spatially fixed paths, which is equivalent to the energy release rate only for a stationary crack in a solid under dynamic motion.

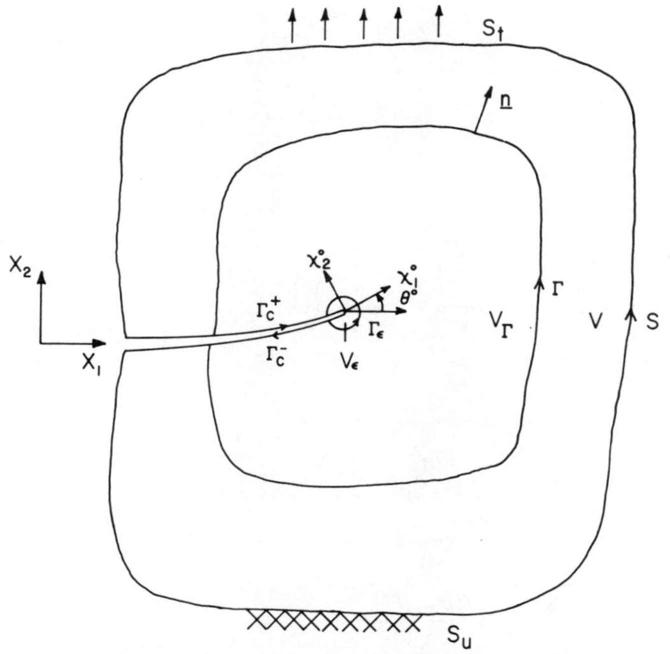

Figure 9. Contours for path-independent integrals.

We assume that either the crack is not kinking or is not abruptly changing the direction and velocity of propagation, i.e., the crack propagation is more or less self-similar. This condition is more precisely stated as follows. Let the vector of crack velocity at time t_1 be $\boldsymbol{C}(t_1)$ and at $t_2 = t_1 + \mathrm{d}t$ be $\boldsymbol{C}(t_2)$. The vector $\boldsymbol{C}(t_2)$ may be decomposed into components $C_t(t_2)$ and $C_n(t_2)$ along and normal to vector $\boldsymbol{C}(t_1)$, respectively. Thus, we consider situations wherein, $|\boldsymbol{C}(t_2)| \simeq |\boldsymbol{C}(t_1)|$ and $C_t(t_2) \gg C_n(t_2)$.

In this case, the energy release rate G per unit of crack-extension is given by (7) and (8). It is noted that (8) is expressed in terms of the global cartesian coordinate system X_i shown in Fig. 9.

The components of the path independent integral which are equivalent to the components of energy release rate were derived in [12] *:

$$J_k' = \lim_{\epsilon\to 0}\int_{\Gamma_\epsilon}[(W+K)n_k - T_i u_{i,k}]\mathrm{d}S$$

$$= \lim_{\epsilon\to 0}\int_{\Gamma+\Gamma_c}[(W+K)n_k - T_i u_{i,k}]\mathrm{d}S + \int_{V_\Gamma - V_\epsilon}[\rho\ddot{u}_i u_{i,k} - \rho\dot{u}_i\dot{u}_{i,k}]\mathrm{d}V. \tag{20}$$

The symbols Γ, Γ_ϵ, V_ϵ, V_Γ, and Γ_c are explained in Fig. 9. Note that the far-field contour Γ is fixed in space. The relations of the various integrals to an energy-release rate are given by

$$G = \frac{C_k}{C}J_k' = \frac{C_k}{C}\left[\hat{J}_k + \int_{\Gamma_\epsilon}Kn_k\mathrm{d}S\right] = \frac{C_k}{C}\left[J_k + 2\int_{\Gamma_\epsilon}Kn_k\mathrm{d}S\right]. \tag{21}$$

For a crack that does not propagate under dynamic loading, all the integrals, J_k, $\hat{J}_k$, J_k are directly related, as seen from (21), to the energy-release rate at incipient crack propagation. This is so because, for a non-propagating crack, the kinetic energy is

* These are components in X_k system (see Fig. 1).

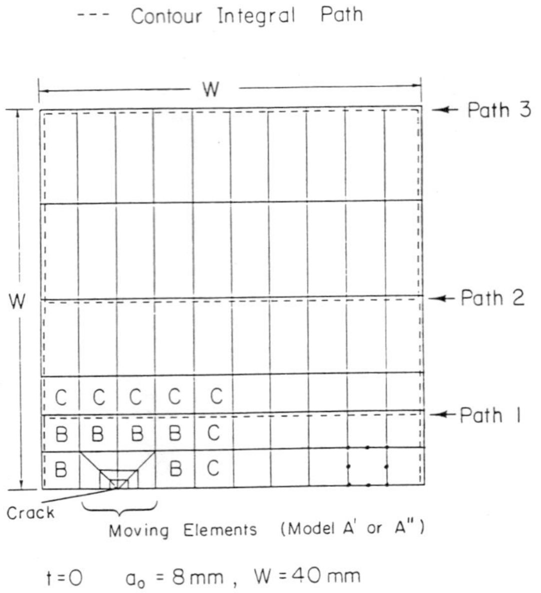

Figure 10. Finite element mesh at time $t = 0$, for a centrally cracked square plate, and contour integral paths.

non-singular and hence $\int_{\Gamma_\epsilon} K n_k \mathrm{d}S \to 0$ in the limit. However, during crack-propagation, K is singular near the crack-tip, and $\int_{\Gamma_\epsilon} \mathrm{K} \mathrm{n}_\mathrm{k} \mathrm{dS}$ remains finite in the limit. Using the asymptotic solutions for elastodynamic crack propagation, given in (4) and (5), the relations between the various path-independent integrals and the instantaneous stress-intensity factors were derived in [12]. The agreement of (20) with that in [48] for the steady-state case has been shown in [49]. *

5.2. Numerical studies of the use of path-independent integrals

A numerical study [50] using the moving singular element procedure indicates that the path-independent integrals J', $\hat{J}$, and J, give distinctly different numerical results, as may be expected from a theoretical point of view (see (21)). Although the moving singular element procedure gives highly accurate solutions, especially for the detailed stress distribution near the propagating crack-tip, this procedure may be difficult to apply by users of general-purpose finite element codes, because of its sophistication. From this point of view, a study was undertaken [51] to use the path-independent integrals along with less-sophisticated finite elements such as isoparametric elements indicated in Fig. 4. The moving-mesh procedure together with regular isoparametric elements (model A″) near the crack-tip gave results in close agreement with those obtained by using the singular element (model A), while the results using quarter-point singular elements (model A′) were considerably different than those from using the singular element for the case of $C = 0.6C_s$. Thus, the model A' can be applied only to slow speed crack propagation, i.e. $0 \leqslant C \leqslant 0.3C_s$.

In the following an example problem of symmetrical dynamic propagation from the tips of a central crack in a square plate is presented. This problem may be considered to be similar to that treated analytically in [52], except that [52] treated an infinite body with

* For a comprehensive survey of recent work on path-independent integrals in inelastic and dynamic fracture see [55].

Table 1. Comparison of path independent integrals and their path independency ($C = 0.6C_s$)

Time	Integral [N/m]	Isoparametric elements (model A″)			Singular element (model A)	Difference (model A″-model A)
		Path 1	Path 2	Path 3	Path 3	Path 3
$t = 20\Delta t$	J_1'	19.35	20.25	20.07	19.95	+0.6%
$(a/w = 0.3)$	$\hat{J}_1$	19.35	20.25	20.07	22.34	−10.2%
	J_1	19.35	20.25	20.07	24.74	−18.9%
$t = 40\Delta t$	J_1'	23.21	23.12	22.95	23.09	−0.6%
$(a/w = 0.4)$	$\hat{J}_1$	23.21	23.12	22.95	25.87	−11.3%
	J_1	23.21	23.12	22.95	28.65	−19.9%
$t = 60\Delta t$	J_1'	28.05	27.99	27.98	20.08	+3.3%
$(a/w = 0.5)$	$\hat{J}_1$	28.05	27.99	27.98	30.33	−7.6%
	J_1	28.05	27.99	27.98	33.58	−16.7%

a crack of zero initial length. Figure 10 shows the finite element breakdown for the initial configuration $t = 0$. A time-independent tensile stress acts at the edge of plate parallel to the crack axis. Three contour paths are considered as shown in Fig. 10.

The results obtained from the models A″ and A are compared in Table 1 for different instants of time and different far-field paths. Reference to Table 1 shows that, while the model A (eigenfunction singular-element) gives distinctly different values for the J', $\hat{J}$, and J integrals (see Fig. 7), the model A″ (non-singular isoparametric elements) gives exactly the same values for the three different integrals throughout the crack propagation history. This is due to the fact that, since the model A″ does not model the singularity in the kinetic energy density K, the limit of $\int_{\Gamma_\epsilon} K n_1 \mathrm{d}S$ in (21) tends to zero. For the model A, however, this singularity was correctly incorporated; thus $\int_{\Gamma_\epsilon} K n_1 \mathrm{d}S$ remains finite. Table 1 also indicates the path independency of the integral values obtained by the moving isoparametric elements. Noting that the relations between each of the three integrals J', $\hat{J}$, and J on the one hand, and the K-factors on the other, are distinctly different from one another [12], it may be seen from Table 1 that the use of the J' integral (20) and its relation to K-factors as given in (11), in conjunction with the use of ordinary (non-singular) isoparametric elements near the propagating crack tip constitutes a computational procedure of sufficient accuracy to analyze fast elastodynamic crack propagation. This assertion was confirmed also in the case of mixed-mode dynamic crack propagation in [53], wherein the path-independent integrals J_k' were used in conjunction with: (i) a moving regular-isoparametric finite-element mesh and (ii) a mode-decomposition method [53,54] to evaluate the mixed mode K-factors.

Acknowledgements

The results presented herein were obtained during the course of investigations supported by the US Office of Naval Research under contract N00014-78-C-0636. The authors gratefully acknowledge this support as well as the encouragement received from Drs Y. Rajapakse and A. Kushner. It is a pleasure to thank Ms J. Webb for her tireless efforts in preparing this typescript.

References

[1] G.T. Hahn and M.F. Kanninen, in *Fracture*, Vol. 1, ICF4, Waterloo (1977) 193–199.
[2] A.S. Kobayashi, *Experimental Mechanics* 23, No. 3 (1983) 338–347.

[3] S.N. Atluri and T. Nishioka, in *Unification of Finite Element Methods*. H. Kardestuncer, ed. North-Holland Press (1984) 65–96.
[4] M.F. Kanninen, in *Numerical Methods in Fracture Mechanics*, D.R.J. Owen and A.R. Luxmoore, eds. Pineridge Press (1978) 612–634.
[5] B.R. Baker, *Journal of Applied Mechanics* 29 (1962) 449–454.
[6] G.C. Sih, R.S. Ravera and G.T. Embley, *International Journal of Solids and Structures* 8 (1976) 977–993.
[7] R.J. Nuismer, Jr. and J.D. Achenbach, *Journal of the Mechanics and Physics of Solids* 20 (1972) 203.
[8] E.P. Chen and G.C. Sih, in *Elastodynamic Crack Problems*, G.C. Sih, ed. Noordhoff International Publishing (1977) 1–58.
[9] J.R. Rice, in *Fracture*, *Vol. II*, H. Liebowitz, ed. Academic Press (1968) 191–311.
[10] L.B. Freund and R.J. Clifton, *Journal of Elasticity* 4 (1974) 293.
[11] J.D. Achenbach and Z.P. Bazant, *Journal of Applied Mechanics* 42 (1975) 183.
[12] T. Nishioka and S.N. Atluri, *Engineering Fracture Mechanics* 18 No. 1 (1983) 1–22.
[13] R.B. Freund, *Journal of the Mechanics and Physics of Solids* 21 (1973) 47–61.
[14] B.V. Kostrov, *International Journal of Fracture* 11, No. 1 (1975) 47–56.
[15] L.R.F. Rose, *International Journal of Fracture* 12, No. 6 (1976) 799–813.
[16] S.N. Atluri, *Engineering Fracture Mechanics* 16, No. 3 (1982) 341–364.
[17] A.S. Kobayashi, A.F. Emery, and S. Mall, *Experimental Mechanics* 16, No. 9 (1976) 231–328.
[18] Y. Ando, G. Yagawa, and Y. Sakai, *Transactions of the 3rd International Conference on SMIRT*, London Paper L 7/1 (1975).
[19] P.N.R. Keegstra, *Journal of the Institution of Nuclear Engineers* 17, No. 4 (1976) 89–96.
[20] P.N.R. Keegstra, J.L. Head and C.E. Turner, in *Numerical Methods in Fracture Mechanics*, A.R. Luxmoore and D.R.J. Owen, eds. University College Swansea (1978) 634–647.
[21] G. Yagawa, Y. Sakai and Y. Ando, in *Fast Fracture and Crack Arrest*. G.T. Hahn and M.F. Kanninen, eds. ASTM STP 627 (1977) 109–122.
[22] J.F. Malluck and W.W. King, in *Numerical Methods in Fracture Mechanics*. A.R. Luxmoore and D.J. Owen, eds. University College Swansea (1978) 648–659.
[23] G. Rydholm, B. Fredriksson, and F. Nilsson, in *Numerical Methods in Fracture Mechanics*. A.R. Luxmoore and D.R.J. Owen, eds. University College Swansea (1978) 660–672.
[24] A.S. Kobayashi, S. Mall, Y. Urabe, and A.F. Emery, *Numerical Methods in Fracture Mechanics*, op cit, 673–684.
[25] S. Mall and J. Luz, *International Journal of Fracture* 16 (1980) R33–R36.
[26] J.F. Malluck and W.W. King, in *Crack Arrest Methodology and Applications*, G.T. Hahn and M.F. Kanninen, eds. ASTM STP 711 (1980) 38–53.
[27] L. Hodulak, A.S. Kobayashi, and A.F. Emery, in *Fracture Mechanics* P.C. Paris, ed. ASTM STP 700 (1980) 174–188.
[28] Z. Bazant, J.L. Glazik Jr. and J.D. Achenbach, *Computers & Structures* 8 (1978) 193–198.
[29] J.A. Aberson, J.M. Anderson, and W.W. King, in *Fast Fracture and Crack Arrest*, op cit, 123–134.
[30] J.A. Aberson, J.M. Anderson, and W.W. King, in *Elastodynamic Crack Problems*. G.C. Sih, ed. Noordhoff (1977) 249–294.
[31] M.L. Williams, *Journal of Applied Mechanics* 24, No. 1 (1957) 109–114.
[32] W.W. King and J.F. Malluck, *International Journal of Fracture* 14 (1978) R7–R11.
[33] C. Patterson and M.C. Oldale, in *Stability Problems in Engineering Structures and Components*. T.H. Richards and P. Stanley, eds. Applied Science Pub. (1979) 281–296.
[34] C. Patterson and M.C. Oldale, in *Advances in Fracture Research* 5, ICF5 (1981) 2225–2232.
[35] T. Nishioka, R.B. Stonesifer and S.N. Atluri, *Engineering Fracture Mechanics* 15, No. 1–2 (1981) 205–218.
[36] S. Aoki, K. Kishimoto, H. Kondo and M. Sakata, *International Journal of Fracture* 14, No 1 (1978) 59–68.
[37] K. Kishimoto, S. Aoki and M. Sakata, *International Journal of Fracture* 16, No. 2 (1980) 3–13.
[38] S.N. Atluri, T. Nishioka and M. Nakagaki, in *Nonlinear and Dynamic Fracture Mechanics*. N. Perrone and S.N. Atluri, eds. ASME AMD Vol. 35 (1979) 37–66.
[39] T. Nishioka and S.N. Atluri, *Journal of Applied Mechanics* 47 (1980) 570–576 and 577–582.
[40] J.F. Malluck, "Crack Propagation in Finite Bodies", Ph.D. thesis, Georgia Institute of Technology (1976).
[41] T. Nishioka, R.B. Stonesifer and S.N. Atluri, in *Numerical Methods in Fracture Mechanics*. D.R.J. Owen and A.R. Luxmoore, eds. Pineridge Press (1980) 489–507.
[42] J.F. Kalthoff, J. Beinert and S. Winkler, in *Fast Fracture and Crack Arrest*. G.T. Hahn et al., eds. ASTM STP 627 (1977) 161–176.
[43] T. Nishioka and S.N. Atluri, *Engineering Fracture Mechanics* 16, No. 3 (1982) 303–332.
[44] S.N. Atluri and T. Nishioka, in *Advances in Fracture Research* 5, ICF5 (1981) 2163–2170.
[45] H.D. Bui, in *Fracture 3*, ICF4, Waterloo (1979) 91–95.
[46] K. Kishimoto, S. Aoki and M. Sakata, *Engineering Fracture Mechanics* 13 (1980) 841–850.
[47] K. Kishimoto, S. Aoki and M. Sakata, *Engineering Fracture Mechanics* 13, No. 2 (1980) 387–394.

[48] G.C. Sih, in *Inelastic Behavior of Solids*. M.F. Kanninen et al., eds. McGraw-Hill (1970) 607–633.
[49] S.N. Atluri and T. Nishioka, in *Fracture Mechanics Technology Applied to Material Evaluation and Structure Design*. G.C. Sih et al., eds. Martinus Nijhoff Publishers (1983) 327–340.
[50] T. Nishioka and S.N. Atluri, *Engineering Fracture Mechanics* 18, No. 1 (1983) 23–33.
[51] T. Nishioka and S.N. Atluri, *AIAA Journal* 22, No. 3 (1984) 409–414.
[52] K.B. Broberg, *Arkiv fur Fysik*, Band 18 (1960) 159–192.
[53] H. Ishikawa, H. Kitagawa, and H. Okamura, in *Proceedings of Third International Conference of Mechanical Behaviors of Materials* 3 (1979) 447–455.
[54] T. Nishioka and S.N. Atluri, "On the Computation of Mixed-Mode *K*-Factors for a Dynamically Propagating Crack, Using Path-Independent Integrals J'''", *Engineering Fracture Mechanics* (To appear).
[55] S.N. Atluri and T. Nishioka, "On Path-Independent Integrals in the Mechanics of Elastic, Inelastic, and Dynamic Fracture"; Special Issue in Honor of Prof. G.R. Irwin, Journal of the Aeronautical Society of India, Nov. 1984.

Résumé

Le mémoire fournit des synthèses des études récentes relatives à la modélisation numérique de la propagation de fissures dynamiques. On examine, à la fois, le maillage "stationaire" et le maillage "mobile" utilisés dans les procédures d'éléments finis. On présente des procédures simples utilisant un maillage "mobile" d'éléments conventionnels isoparamétriques utilisé avec certaines intégrales indépendantes du parcours, en vue d'évaluer les facteurs d'intensité de contrainte dans le cas d'une fissure en cours de propagation dynamique.

International Journal of Fracture 27 (1985) 263–276.

Application of dynamic shear crack models to the study of the earthquake faulting process

SHAMITA DAS

Lamont-Doherty Geological Observatory of Columbia University, Palisades, NY 10964, USA

(Received September 1, 1984; in revised form November 2, 1984)

Abstract

The earthquake faulting process can be modelled as propagating two or three dimensional shear cracks. Since analytical methods cannot be used to study finite cracks, several numerical approaches to this problem have been taken, of which the numerical boundary integral method is here described in detail. The process of crack growth can be determined by the use of a fracture criterion. Since the material properties such as fracture strength and frictional properties vary over the surfaces of real faults, two types of complex faulting models – the "barrier" model and the "asperity" model – are discussed in the context of seismology.

1. Introduction

Information on the earthquake faulting process is obtained from analyzing the seismic waves that are radiated when the two sides of a fault slip past one another at average velocities on the order of a few metres per second. These waves are recorded by seismometers located at or very near the surface of the earth. The ground motion at any site is a space-time convolution of the time history of slip on the disturbed area of the fault with the impulse response of the medium through which the waves propagate from the fault area to the receiver. This relation is known as the representation theorem for seismic sources [1] developed using the representation theorems for elastodynamics [2]. A major goal of earthquake seismology is to solve the inverse problem of determining the time history of slip from a study of these recorded seismograms. Unfortunately, this problem is mathematically so unstable that it is impossible to solve the inverse problem uniquely by a trial-and-error parameter fitting of models to the observed ground motion. There is no way of overcoming the instability other than by constraining the class of permissible models based on some physical information, independent of seismic observations. Fracture mechanics theories can provide a unique source of such information.

Considerable progress has recently been made in modelling crustal earthquake as propagating shear cracks, based on theories originally developed for the study of tensile cracks. It is possible, at least in theory, to numerically solve the forward problem for a realistic case where the material properties, such as fracture strength and frictional properties on a fault are spatially variable along with a realistic Earth model. In practice, one is limited by the fact that the initial or the ensuing physical conditions on the fault – which is usually at depths of the order of tens of kilometers in the earth – before and during the earthquake are not known, so that important parameters necessary for modelling are unknown. Seismologists are currently studying the forward problem for simple cases to gain insight into the physical processes that occur on or in the vicinity of propagating faults. In this respect, the crack propagation problem as applied to seismology is rather different and more difficult than fracture problems in engineering, where the stress environment and material properties of the medium are usually well known.

The difference in scales between the seismological problem and the fracture mechanical problems might seem at first to demand different approaches to each of these classes of problems. In fracture mechanics, the dimensions of the cracks under consideration are usually a few microns to at most tens of metres in man-made structures. Earthquakes result from shear cracks which propagate hundreds of kilometers – the 1960 Chilean earthquake had a fault length of about 800 km – at average speeds of about 3 km/sec. But in reality, the same theories can be applied to the propagation of microcracks as well as to these gigantic cracks.

In current seismic source modelling, the faulting process is described in terms of mode II or mode III cracks or as three-dimensional shear cracks. For simplicity, the medium is taken to be infinite, homogeneous and linearly elastic everywhere except near the crack edge. This paper reviews the types of models that are currently being used in interpreting seismograms. A very brief outline of the current literature follows next.

2. Summary of current literature

In looking back on the development of crack models, it is seen that the theories for extending mode I and mode III cracks developed first and the mode II crack solutions followed later. Here we shall discuss dynamic shear crack models only.

The first fracture mechanical theories for studying dynamic shear cracks were developed about two decades ago [3,4]. The slip and stress field for a self-similar circular shear crack in an infinite, homogeneous elastic medium which initiates at a point and grows at a speed less than the shear wave speed β were found analytically [3]. The solution for a self-similar mode III crack which grows at a variable velocity was determined using an analytical boundary integral equation method. It was found that mode III cracks which propagate under the fracture criterion that the rate of energy flux into the crack tip is constant (dynamic form of the Griffith criterion) gradually accelerate from zero to the shear wave speed β [4]. Since then, the solutions for an expanding, self-similar elliptical shear crack [5] and for mode III cracks with friction [6] have been found. It was shown [7] that for mode II cracks lacking cohesion, the fracture speed may be as high as the compressional wave speed α of the medium. Analytical expressions have been found [8] for the stress-intensity factor for mode II cracks with variable loading.

Analytical methods cannot be used to find the solutions for finite, dynamic cracks, since the multiple integrals resulting from the repeated diffraction of waves at the crack edge cannot be obtained in closed form even in the simplest cases [4]. Hence, since the late '60's most studies of dynamic shear crack propagation have focused on numerical methods. A numerical singular integral equation technique to study finite mode II and mode III cracks which expand at known constant speeds and may stop was developed and used to find the time history of slip on such cracks [9]. A finite difference technique and a "slip-weakening" fracture criterion [10] was used [11] to show that even for cracks with cohesion, the maximum crack tip speed may be as high as the compressional wave speed α for mode II cracks. This numerical result differs from the result reached from analytical considerations [12] that the maximum permissible speed for mode II cracks is the Rayleigh wave speed of the medium and is a result which is still the subject of discussion in seismology. A numerical boundary integral technique was developed for mode II and mode III cracks [13], by a method that is also applicable to mode I cracks, and used with a "critical stress level" fracture criterion in which a grid point just ahead of the crack tip breaks when the average stress over the grid reaches some critical value [13] as well as with a "slip-weakening" fracture criterion [14].

In 1976, a numerical solution for a circular crack which grows at a given speed and stops was developed and used to find the relation between the slip and stress drop on circular faults [15]. Since then, several numerical studies of three-dimensional crack problems have been performed, both for propagation at known velocity [16–19] and for spontaneous propagation using a fracture criterion [20–22]. These studies provided relationships between slip and stress drop on faults other than circular ones and confirmed in three dimensions the earlier two-dimensional results such as finding that the maximum in-plane crack speed could exceed the Rayleigh wave speed for weak faults. Later on, we shall expand on the results of some of these studies.

3. Heterogeneous dynamic faulting models

The faulting process depends on the "dynamic stress drop" defined as the initial stress minus the dynamic frictional stress and on the critical stress relative to the initial stress required for sliding. In nature both these quantities probably vary over the fault surface. Two broad classes of simple dynamic models are currently used to describe the non-uniform faulting processes occurring during earthquakes. In one, called the "barrier" model, the relative critical stress varies over the fault [23]. In the other, called the "asperity" model, the dynamic stress drop varies over the fault [24,25]. The results for the barrier model will be shown in a later section.

The spontaneous, dynamic problem for the case when the critical stress and dynamic stress drop both vary [21] was studied using a finite element method and the slip-weakening fracture criterion. In this model, the regions of high critical stress are regions where the dynamic stress drop vanishes and regions of high dynamic stress drop are regions where the critical stress needed for sliding is small. It was shown that regions of high fracture speed are regions where the dynamic stress drop is large and the critical stress is small and these are also regions of high slip velocity. It was also shown that in these regions the fracture speed can exceed the shear wave speed, though the *average* speed for the whole fault is still lower than the shear wave speed. Recent evidence from the 1979 Imperial Valley, California earthquake strongly suggests that locally the fracture may have propagated at a speed higher than the shear wave speed [26].

4. The numerical three-dimensional boundary integral method

We next describe the three-dimensional boundary integral method [19] in detail mainly as a vehicle for demonstrating the formulation of the dynamic crack propagation problem in three dimensions. We describe the problem for the case when the velocity of the crack edge is assumed to be known. This method can be applied to the study of dynamic tension cracks in three dimensions as well.

The geometry of the shear crack in an infinite, homogeneous, elastic body is shown in Fig. 1. Let $x_3 = 0$ be the plane over which the crack spreads out radially starting from point 0. The body is initially stressed, with $\sigma_{ij}^0(x_k)$ denoting the distribution of stress. In the earth the stress components, $\sigma_{ij}^0(x_k)$ vary with time over long periods (decades) but during the earthquake event, σ_{ij}^0 may be assumed independent of time.

The stress on the fault plane $x_3 = 0$ can be separated into the normal stress σ_{33}^0 and shear stresses σ_{13}^0 and σ_{23}^0. Let the initial shear stress be increased so that the crack extends dynamically. The normal stress σ_{33}^0 over the plane $x_3 = 0$ remains constant throughout the fracture process as long as the crack remains planar and no crack opening occurs. This assumption is valid in the earth due to the very large compressive stresses created by gravity at depths of tens of kilometers. The crack extends by generating a displacement discontinuity over the surface $x_3 = 0$. The shear stress on the broken surface of the crack

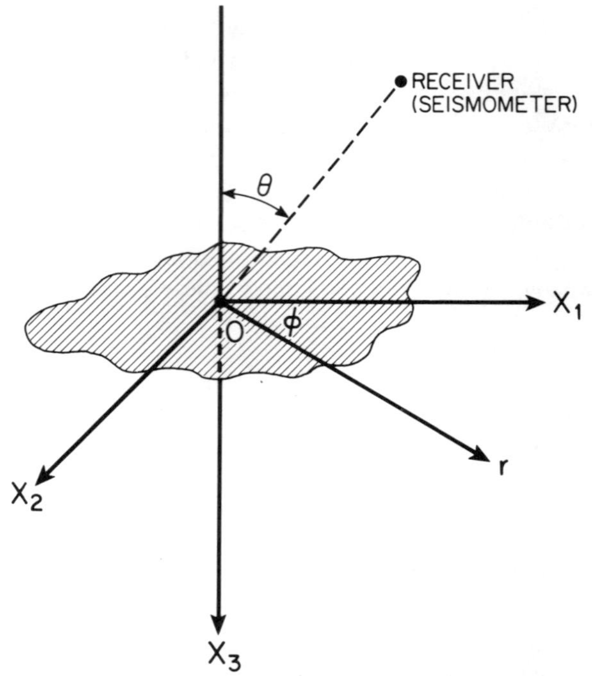

Figure 1. Schematic representation of crack geometry. $x_3 = 0$ is the fault plane.

plane can be zero if there is complete stress release or may be taken to be some finite value. In the latter case, the dynamic frictional stress may be given by $\mu_d \sigma_{33}^0$, where μ_d is the dynamic coefficient of friction which may be a function of x_1 and x_2. Let the incremental stresses due to the displacements $\boldsymbol{u}$ from the initial configuration be τ_{ij} so that the total stress at any time after slip initiation is $\sigma_{ij} = \sigma_{ij}^0 + \tau_{ij}$. For initial conditions one assumes displacements and velocities vanish everywhere. If the initial displacements are non-zero, i.e., instead of a crack starting from a point we have a pre-existing finite crack that grows, the dynamic part of the solution for displacements will be unaffected by the initial displacements; for, one can first solve the problem for zero displacements and then add the initial displacement field to the dynamic solution to obtain the final displacement. Note however that the radiated waves (i.e., seismograms) are only sensitive to changes of displacement and hence are unaffected by the initial solution.

Due to the symmetry properties of stress and displacement [19] across the $x_3 = 0$, it is sufficient to solve the problem in a half-space bounded by the crack plane $x_3 = 0$. Because the normal stress does not change in this problem, $\tau_{33} \equiv 0$. The integral equation for the displacements may be written as [19,27]

$$u_1(x_1, x_2, 0, t) = \iiint_S [g_{11}(y_1 - x_1, y_2 - x_2, 0, 0, t - s)\tau_{13}(y_1, y_2, 0, s) + g_{12}(\ldots)\tau_{23}(\ldots)]\,\mathrm{d}y_1\mathrm{d}y_2\mathrm{d}s$$

and (1)

$$u_2(x_1, x_2, 0, t) = \iiint_S [g_{21}(\ldots)\tau_{13}(\ldots) + g_{22}(\ldots)\tau_{23}(\ldots)]\,\mathrm{d}y_1\mathrm{d}y_2\mathrm{d}s,$$

where $g_{ij}(\boldsymbol{x}, t; \boldsymbol{y}, s)$ are half-space Green's functions i.e., the displacement in the i

direction at the point $(x_1, x_2, 0)$ at time t due to an unit impulse acting in the j direction at $(y_1, y_2, 0)$ at time s. The region of integration extends over the disturbed region on the $x_3 = 0$ plane, given by $\alpha^2(t-s)^2 - (y_1 - x_1)^2 - (y_2 - x_2^2) \geqslant 0$, $t \geqslant s \geqslant 0$, α being the compressional wave speed of the medium. If τ_{13} and τ_{23} were known everywhere within the slip region one could solve the above integrals to find u_1 and u_2. However, τ_{13} and τ_{23} are the stress changes due to crack propagation on the fracture (S_1) as well as on the unfractured portions (S_2) of $x_3 = 0$. We shall solve the problem when τ_{13} and τ_{23} are known on S_1 but not on S_2. The boundary conditions for the problem are then:

On S_1, τ_{13}, τ_{23} known and equal to the stress drop due to fracture, $\tau_{33} = 0$, (u_1, u_2 and u_3 unknown)
On S_2, $u_1 = 0 = u_2$, $\tau_{33} = 0$ (τ_{13}, τ_{23} and u_3 are unknown).

To determine τ_{13} and τ_{23} in S_2, we use the fact that $u_1 = 0 = u_2$ in S_2 and write (1) in discrete form. Then, by taking terms for which $x_1 = y_1$, $x_2 = y_2$ and $s = t$ on the left-hand side of the equation, we find

$$\begin{aligned} F_{12}\tau_{23}(x_1, x_2, 0, t) + F_{11}\tau_{13}(x_1, x_2, 0, t) &= -\sum_{\substack{\text{sum over}\\ x_1, x_2, s}}\sum\sum [g_{12}(\ldots)\tau_{23}(\ldots) + g_{11}(\ldots)\tau_{13}(\ldots)] \\ &= L, \quad \text{say} \end{aligned}$$

$$\begin{aligned} F_{22}\tau_{23}(x_1, x_2, 0, t) + F_{21}\tau_{13}(x_1, x_2, 0, t) &= -\sum_{\substack{\text{sum over}\\ x_1, x_2, s}}\sum\sum [g_{22}(\ldots)\tau_{23}(\ldots) + g_{21}(\ldots)\tau_{13}(\ldots)] \\ &= M, \quad \text{say,} \end{aligned}$$

where F_{ij} are the known values of the Green's function g_{ij} averaged over the grid system centered at the origin of the coordinate system. The summation extends over every point of the cone S except the vertex. Due to the symmetries in the Green's functions, it follows that:

$$\begin{aligned} F_{ij} &= 0, \quad \text{if } i \neq j \\ &= 0, \quad \text{if } i = j. \end{aligned}$$

Also, $F_{11} = F_{22}$. Hence, the unknown stresses are determined as

$$\tau_{13}(x_1, x_2, 0, t) = \frac{L}{F_{11}},$$

and

$$\tau_{23}(x_1, x_2, 0, t) = \frac{M}{F_{11}}.$$

Once τ_{13} and τ_{23} are known everywhere in S_1 one uses the discrete form of (1) to find u_1 and u_2 in S_1. The required Green's functions for this problem were determined analytically for a Poisson solid ($\lambda = \mu$) [28] and for general Lamé parameters λ and μ [29].

As a test of the method, the numerical results are compared in Fig. 2 with analytical results [3] for a self-similar circular shear crack that expands at a constant speed of $\alpha/2$, where two grid-sizes were used to assess the accuracy and stability of the method.

It is appropriate here to briefly discuss the advantages and disadvantages of different numerical methods that are being used at present to study dynamic shear crack problems. The finite difference and finite elements methods are extremely well understood and well-documented and amenable to generalization, e.g., the dynamic crack problem can be

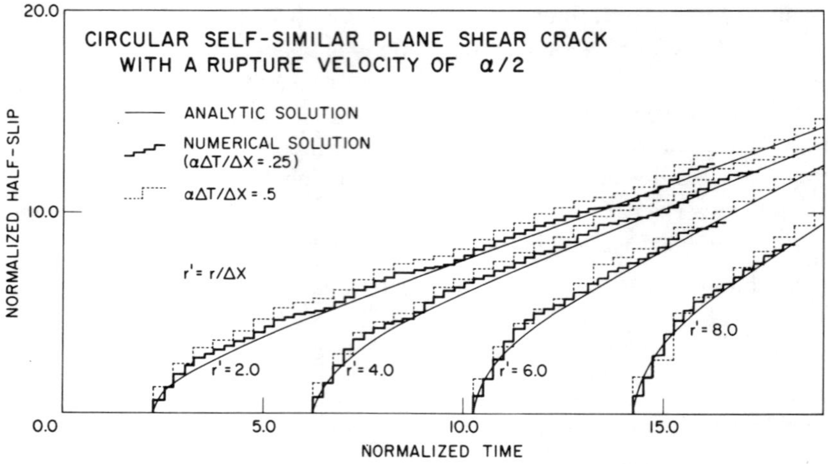

Figure 2. Comparison of analytical and numerical half-slip for a circular shear crack propagating at a speed of $\alpha/2$. The analytic solution [3] is given by the continuous line. The numerical solution is given by step-like lines for $\alpha\Delta t/\Delta x = 0.5$ (dotted) and 0.25 (solid). Normalized time is $\alpha t/\Delta x$. The normalization factor for the half-slip is $\Delta x \Delta\tau/3\mu$. μ is the modulus of rigidity, and Δx and Δt are the spatial and temporal grid sizes. $\Delta\tau$ is the stress drop on the crack and r is the instantaneous crack radius.

solved in a layered half-space using these methods. They use extremely large space and time in computers and are thus limited by these considerations to how fine a grid one can use. In the integral equation techniques, the problem reduces to calculating quantities only on the crack plane; it is thus inherently more accurate, results in large savings in computer space and time and does not suffer from dispersion of short wavelength components. This method in far less amenable, however, to modifications for the introduction of free-surfaces and heterogeneous media. The difference between the singular integral equation method and the boundary integral method lies in the fact that in the former the stress is written as a space-time convolution of a slip on the fault with the second derivatives of whole space Green's function, whereas in the latter case the displacement is a space-time convolution of the half-space Green's function with the stress change on the fault. The integrand in the latter case has a lower order singularity than that in the former case and the regions over which the integration extends is different in the two formulations. In the boundary integral method, the region of integration is the entire, disturbed region of the crack plane. In the singular integral equation method, the region of integration is confined only to the slipping part of the crack plane. If the crack speed is less than the maximum wave speed of the medium, this results in savings of computer time for "interior" crack problems. However, the situation is reversed for "exterior" crack problems and then the boundary integral method becomes more advantageous in this respect.

5. Spontaneous shear crack propagation

Let us consider a pre-existing crack in an infinite, elastic body and assume for the moment, that the initially applied shear stress has only one component $\sigma_{13} = \sigma_0$, say. A cross-section through the x_1-axis of the initial stress field is shown in Fig. 3a. In the initially broken region of the crack plane, the stress is at a lower level σ_1 and at the crack periphery there is a stress concentration. During fracture, the crack expands and the resulting stress field at some later time is shown qualitatively in Fig. 3b for a crack propagating at less than the Rayleigh wave speed. The stress on the crack, including the initially broken region, is taken as the dynamic frictional stress σ_d. Note that in dynamic

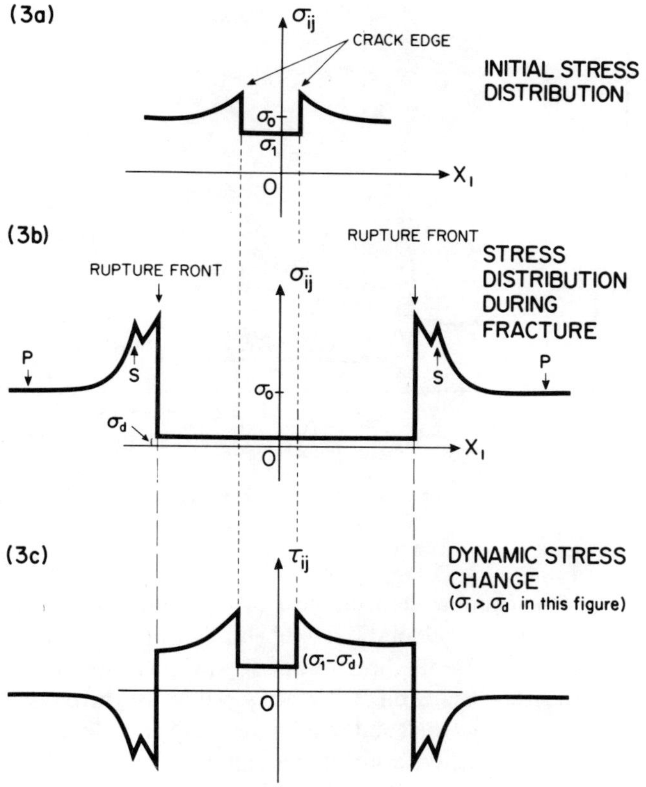

Figure 3. (a) A cross-section along the x_1-axis of the initial stress distribution on a pre-existing crack. (b) Stress distribution during expansion of crack in cross-section. The S waves ahead of the tip (for a crack propagating at a speed less than the Rayleigh wave speed, say) can be seen. The Rayleigh wave speed is 0.9β for a Poisson solid. (c) Dynamic stress change.

problems, this stress concentration depends on the crack tip velocity [30,31]. The stress change during fracture propagation (τ_{ij}) is obtained by subtracting the problem corresponding to Fig. 3b from that of Fig. 3a with the result as shown in Fig. 3c. σ_1 may greater than or less than σ_d. Figure 3c is plotted assuming $\sigma_1 > \sigma_d$, and τ_{ij} on the initially broken crack is then $(\sigma_1 - \sigma_d)$. The difference between the initial shear stress and the shear on the crack during fracture is called the "dynamic stress drop" $\Delta\tau$; $\Delta\tau$ is the same as τ_{ij} on the broken part of the fault and is zero on the unbroken region: *it is the driving force of earthquake faulting.*

Next, let us confine our attention to a particular point well ahead of the crack tip and study the change of stress there as a function of time (Fig. 4). Initially, the shear stress at the point is σ_0. As the crack starts expanding the stress starts increasing there. The details of this increase depends on the manner in which fracture occurs. In the "critical stress level" criterion [13] this point breaks when the stress reaches some critical level, say σ_u. σ_u can be interpreted as the fracture strength of the material in shear, if the fault is propagating through previously unbroken material. In many cases, the two sides of the fault are actually physically broken but are held together by the static frictional stress. In such cases, σ_u is interpreted as the static frictional stress given by $\mu_s\sigma_{33}$, where μ_s is the coefficient of static friction. The stress at this point then drops, either instantaneously, or over some finite time to the dynamic frictional level, σ_d. σ_d is plotted as constant in Fig. 4 but may equally well be variable, in reality. When the crack edge stops due to various reasons discussed later the waves diffracted from the boundary return to interior points

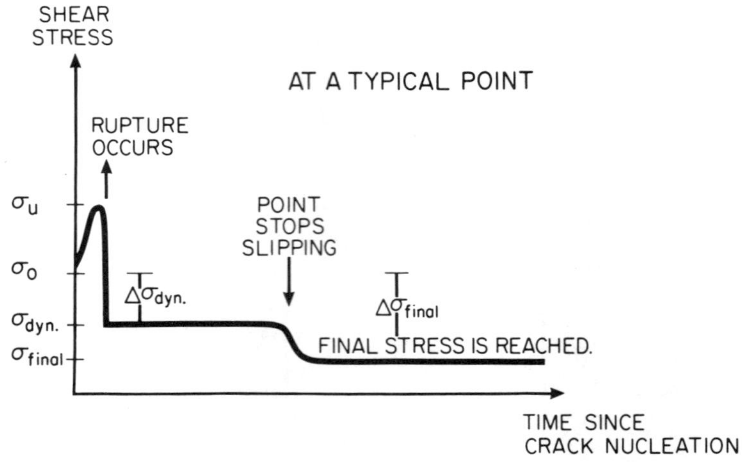

Figure 4. Stress change as a function of time ahead of the crack tip.

carrying with it the information on the finiteness of the fault and cause interior points to stop slipping. In most numerical calculations, the slip "stopping criterion" is that a point is "frozen" when it tries to reverse its direction of slip [15,32]. It is found that points slow down and stop soon after the shear and Rayleigh waves from the closest final crack edge return to it. Thus points closer to the final crack edges slip for shorter times than those in the interior. For such a stopping criterion, the final stress σ_f on the crack as the whole system comes to rest is *lower* than σ_d. The quantity $(\sigma_0 - \sigma_f)$ is termed the "final stress drop" and it is therefore larger than the "dynamic stress drop".

The slip velocity has a (square-root) singularity at the crack edge – a "velocity-intensity" factor can be associated with this singularity – which falls off as one moves into the crack from the edge. The slip, slip velocity, and stress field due to a propagating crack depend linearly on the dynamic stress drop [1,2]. Thus, if the dynamic stress drop is doubled, the slip and slip velocity on the crack and the stress intensity factor at the tip are doubled and so on so that the problem has to be solved only once for a given $\Delta\tau$ and can be scaled for other values of $\Delta\tau$. The crack propagation velocity problem, however, is a non-linear problem. The crack velocity depends only on the quantities $(\sigma_u - \sigma_0)$ and $\Delta\tau$. For each value of these quantities, the calculations have to be redone to find the crack edge position as a function of time. The larger $(\sigma_u - \sigma_0)$ the slower is the fracture speed and vice versa.

Let us next compare the dynamic forms of the Griffith and Irwin criteria. In the dynamic form of Griffith's criterion [4,13] the energy required to create a unit area of crack surface is a constant, i.e. independent of crack speed. In the dynamic form of Irwin's criterion [13], the critical stress intensity factor is a constant. For static problems, these two criteria are identical. For dynamic problems, however, the crack tip speed using these two criteria differ. Figure 5a shows the crack tip position with time for a semi-infinite mode III crack using the Griffith and Irwin criteria and Fig. 5b shows the crack tip speed for the same problem. It is found that the crack with the dynamic Griffith criterion reaches its terminal velocity β faster than that with the dynamic Irwin criterion. The fracture energy release G per unit time decreases linearly from its static value to zero at the terminal crack speed β, for the dynamic Irwin crack but this decrease is non-linear for the Griffith crack. See [33] for further comparison of these two criteria.

Since these fracture criteria are not easily adaptable to numerical computation, the "critical stress level" criterion defined earlier is often used in numerical schemes [13,22].

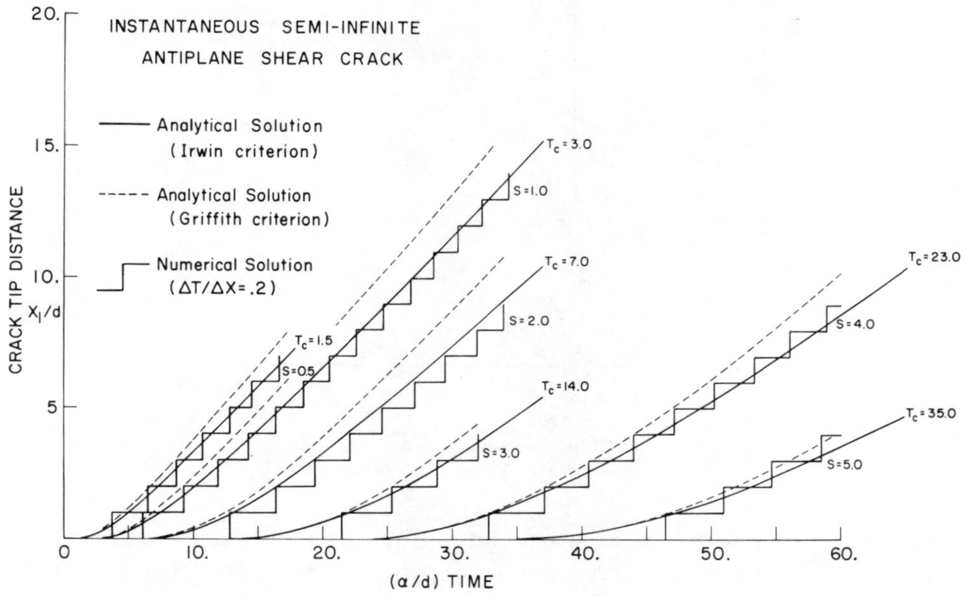

Figure 5(a). Comparison of crack tip position as a function of time for a semi-infinite, instantaneous mode III crack using the dynamic Griffith, the dynamic Irwin and the critical stress level criterion [13]. The shear wave velocity is taken as unity. S is $(\sigma_u - \sigma_0)$/dynamic stress drop. $T_c = \alpha t c/\Delta x$, where t_c is the time the crack-tip waited before extending and Δx and Δt are the spatial and temporal grid sizes.

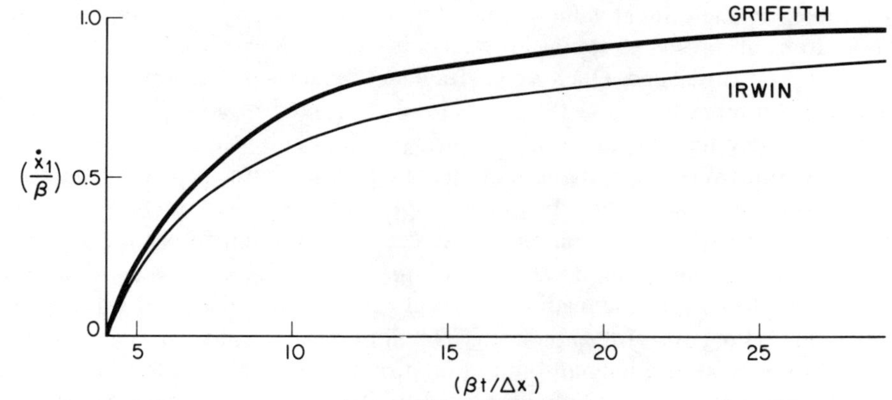

Figure 5(b). Crack tip velocity as a function of normalized time. β is the shear wave velocity, t is time and Δx is the spatial grid size.

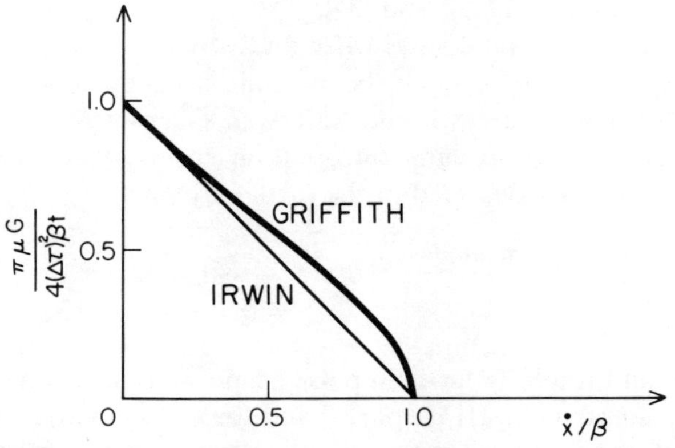

Figure 5(c). Fracture energy G absorbed by the moving mode III crack-tip per unit time as a function of normalized crack speed. μ is the shear modulus, β is the shear wave velocity and t is time.

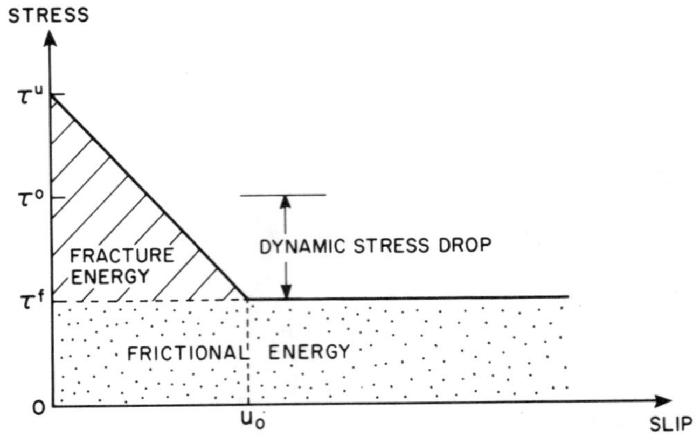

Figure 6. "Slip-weakening" fracture criterion.

The crack tip position with time for a semi-infinite mode III crack under this criterion is also shown in Fig. 5a, and is very close to that found for the dynamic Griffith and dynamic Irwin criteria. Thus, this simple criterion, inspite of its drawback that it is grid size dependent, does give a general idea of the crack tip behaviour.

Let us next discuss the "slip-weakening" fracture criterion. The stress at a point along the crack edge is shown in Fig. 6 as a function of slip. As the shear stress increases from the initial value σ_0 to the critical value σ_u, no slip occurs. Once σ_u is reached, the stress drops from σ_u to σ_d according to some slip weakening law (a linear law is used in Fig. 6) until a critical slip u_0 is reached. Once u_0 is attained, the stress has dropped to the value σ_d and the point continues to slip at this stress level. The hatched area gives the fracture energy G. Results using the critical stress level criterion [20] and using the slip-weakening criterion for three-dimensional problems [21] agree in general character and no significant discrepancy is found between them. For cracks spreading out over weak planes where $\sigma_u - \sigma_0$ is small compared to the dynamic stress drop and constant over the plane, the terminal velocity at the purely mode II parts of the crack edge can be as high as the compressional wave speed α of the medium but is always less than the shear wave speed β at the purely mode III regions of the crack edge for all materials. Thus, on weak planes the cracks will tend to be relatively long and thin. For stronger materials, the terminal velocity for the pure mode II parts of the crack edge are found to be less than the Rayleigh wave speed of the medium and a crack on such a plane would be approximately equidimensional. The behaviour of slip for rectangular faults for various types of boundary conditions has been studied in [17,20] and [34].

An important question in seismology is when and why cracks stop [35,27]. In short the crack stops when the ratio of $(\sigma_u - \sigma_0)$ to the dynamic stress drop becomes too large over an extended area i.e., either the crack encounters an extensively strong region through which it cannot break or it simply runs out of strain energy. It has been suggested that these "arrest mechanisms" are observed in the earth [35].

6. Results for heterogeneous fault models

Barrier model

The slip distribution and resulting far-field pulse shapes were calculated for a fault plane with barriers for a bilateral mode III crack [27] and for a unilateral mode II crack [27,23]. Here I show some of my unpublished results for a three dimensional shear crack. Three types of fault were studied. Type I results when $(\sigma_u - \sigma_0)$ is constant over the fault, Type

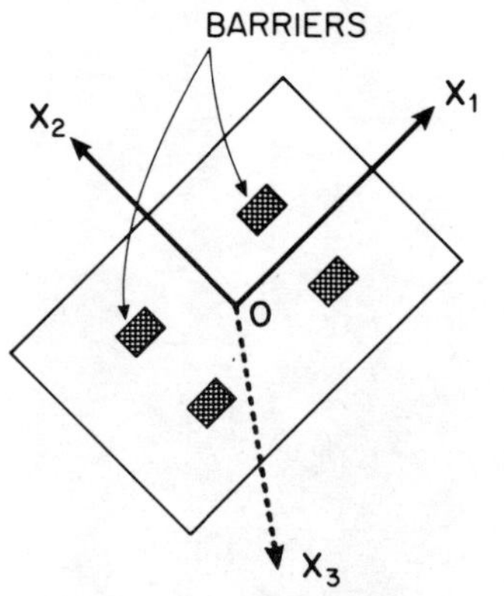

Figure 7. "Barrier" model in 3-D (Type II and Type III cracks).

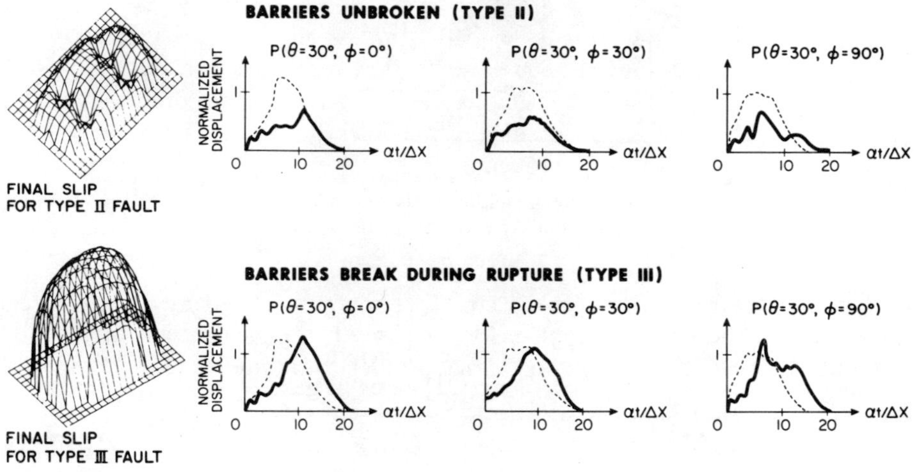

Figure 8. The final slip for Type II and III cracks is shown as perspective plots on the left. The far-field *P* displacement pulse shapes for Type II and Type III cracks are shown on the right. The pulse shape for Type I cracks is superimposed on these figures in dotted lines. θ and ϕ are shown in Fig. 1.

II (Fig. 7) when $(\sigma_u - \sigma_0)$ is so large in the hatched areas (barriers) that those regions remain unbroken and the fracture propagates around them; Type III is obtained when $(\sigma_u - \sigma_0)$ in the hatched areas is large enough that these points do not break at first and the fracture propagates around it but while the fracture and ensuing slip on the fault is still continuing and further loads these regions, they break resulting in a very complex fracture process. Figure 8 shows the final slip for Type II and Type III fractures (the final slip for Type I is very similar to that for Type III) as well as the far-field *P* displacement wave-forms looking along three different directions from the fault. The Type I wave-form (smooth crack-dotted line) is superimposed on the Type II and Type III wave-forms for comparison. Note the difference in the frequency-content and the position in time of the maximum slip. Similar results were found for a scalar analog of Type I and Type II cracks moving at a preassigned velocity [18].

Asperity model

As a very idealized case of a heterogeneous faulting, in which one concentrates attention on the details of the strong regions ("asperities") on a fault fracture, one may study the

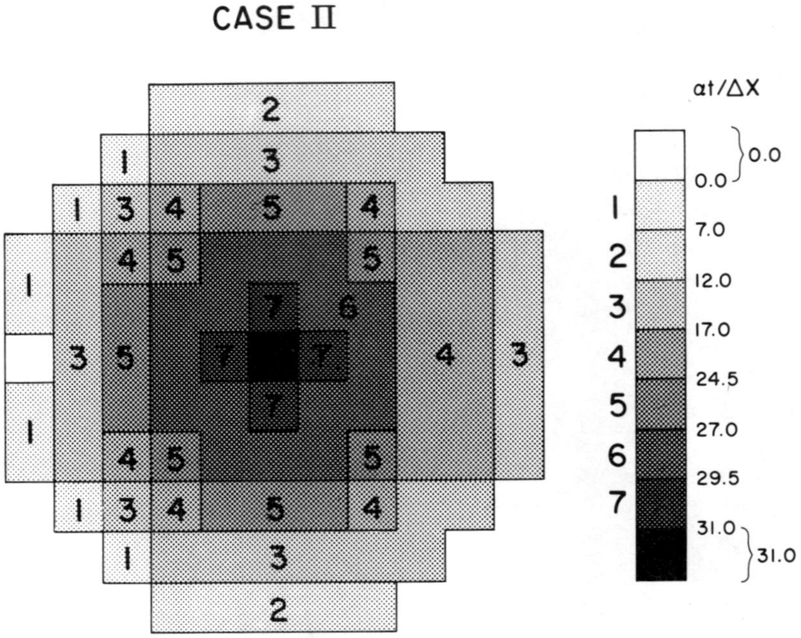

Figure 9. Fracture areas as a function of time for a circular "asperity" to show the "double encircling pincer movement". The blank area is where fracture initiated. The key indicates the different areas fractured during the normalized time-intervals marked. The initial stress is along the x_1 axis.

problem of fracture of an unbroken strong region surrounded by a broken region on a fault. The dynamic fracture problem for the case of a single circular asperity in shear on an infinite fault plane (the "exterior" crack problem) was studied [35]. The initial stress on

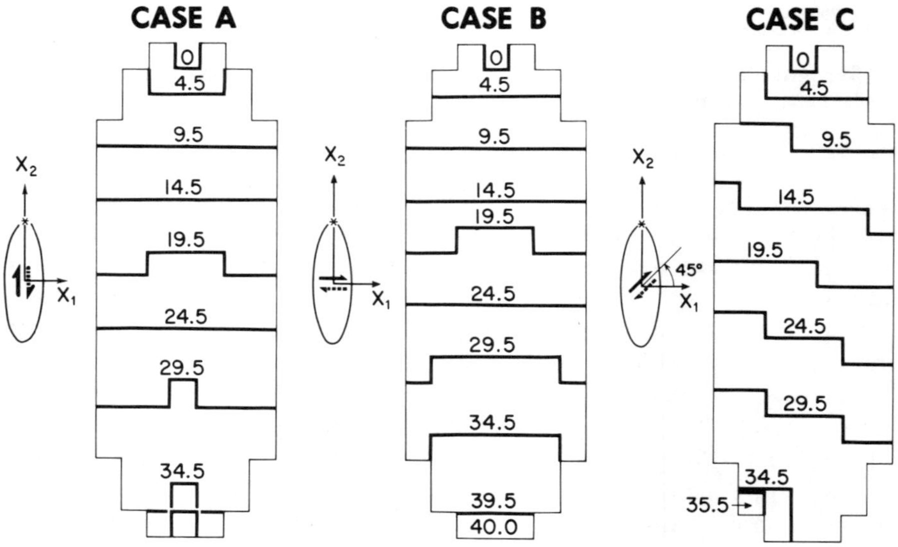

Figure 10. Three cases of elliptical asperity fracture. The ellipse is of major to minor axis ratio of 3:1. Star indicates point of fracture initiation, which is always found to be at or near the end of the major axis. The direction of initial stress is shown in the small inset by each case. The numbers denote the normalized time interval $\alpha t/\Delta x$, where Δx is the spatial grid size. Case A and Case B are primarily mode II and mode III respectively. Case C is a mixed mode problem. For the σ_u used for Case C, the mode II and mode III components propagate at different speeds resulting in the rotation of the propagating wave front.

the circular region was determined using an existing solution for the circular contact problems [37]. It was found as shown in Fig. 9 that for an asperity of constant σ_u the fracture first propagates around the perimeter where the stress is concentrated initially and then propagates inward. The dynamic problem of fracture of an elliptical asperity of constant σ_u was studied [38] to show that the shape of the asperity strongly controls the fracture process. For the elliptical asperity in which the stress concentration is always larger at the end of the major axis than at the end of the minor one, with a σ_u just greater than the stress concentration at the ends of the major axis and where fracture is initiated by relaxing a grid at one end of the major axis, the fracture front sweeps across the asperity from one end of the major axis to the other. Of the several cases studied, three are shown in Fig. 10.

7. Concluding remarks

This review summarizes some of the major results obtained in the last two decades in solving dynamic shear crack propagation problems with a view to applying them to the understanding of the wave-field radiated during an earthquake. The final goal of these studies is to provide insight into the physical processes occurring on the fault during an earthquake from the analyses of seismograms.

Methods for solving very complex numerical models of spontaneous shear cracks propagating on planes of variable strength, variable initial stress and variable stress drop have been developed to simulate shallow faulting in nature. Since little is known about the actual variability of these physical parameters at deeper levels in the earth, these numerical simulations are now being used to study changes in these parameters to understand the effect of such simple variations on the fracture process and the resulting radiated wave field.

Some excellent reviews of related topics can be found in [25,30] and [39–42].

Acknowledgements

I would like to thank Professor W.G. Knauss for a very thorough review which improved the paper. Professor Ben Freund also provided useful comments. Chris Scholz, Barry Raleigh and Keith Evans were the Lamont internal readers and I thank them for their reviews. This work was supported by the National Science Foundation under grant EAR81-20318 and the US Geological Survey under contract USGS-14-08-0001-G-947. Lamont-Doherty Contribution No. 3768.

References

[1] R. Burridge and L. Knopoff, *Bulletin of the Seismological Society of America* 54 (1964) 1875–1888.
[2] A.T. de Hoop, Representation Theorems for the Displacement in an Elastic Solid and their Applications to Elastodynamic Diffraction Theory", D.Sc. thesis, Technische Hogeschool, Delft (1958).
[3] B.V. Kostrov, *Journal of Applied Mathematics and Mechanics* 28 (1964) 1077–1087.
[4] B.V. Kostrov, *Journal of Applied Mathematics and Mechanics* 30 (1966) 1241–1248.
[5] R. Burridge and J.R. Willis, *Proceedings of the Cambridge Philosophical Society* 66 (1969) 443–468.
[6] R. Burridge and G.S. Halliday, *Geophysical Journal of the Royal Astronomical Society* 25 (1971) 261–283.
[7] R. Burridge, *Geophysical Journal of the Royal Astronomical Society* 35 (1979) 439–455.
[8] B.V. Kostrov, *Journal of Applied Mathematics and Mechanics* 38 (1974) 551–560.
[9] R. Burridge, *Philosophical Transactions of the Royal Society of London* A265 (1969) 353–381.
[10] A.C. Palmer and J.R. Rice, *Proceedings of the Royal Society of London* A332 (1973) 527–548.
[11] D.J. Andrews, *Journal of Geophysical Research* 81 (1976a) 5679–5687.
[12] A.F. Fossum and L.B. Freund, *Journal of Geophysical Research* 80 (1975) 3343–3347.
[13] S. Das and K. Aki, *Journal of the Royal Astronomical Society* 50 (1977) 643–668.

[14] D.J. Andrews, *Bulletin of the Seismological Society of America*, submitted (1984).
[15] R. Madariaga, *Bulletin of the Seismological Society of America* 66 (1976) 639–666.
[16] S.M. Day, Three dimensional finite difference simulation of fault dynamics, Final Report SSS-R-80-4295, S-Cubed, La Jolla, CA (1979).
[17] S.M. Day, *Bulletin of the Seismological Society of America* 72 (1982) 705–727.
[18] R.B. Burridge and R. Moon, *Geophysical Journal of the Royal Astronomical Society* 67 (1981) 325–342.
[19] S. Das, *Geophysical Journal of the Royal Astronomical Society* 62 (1980) 591–604.
[20] S. Das, *Geophysical Journal of the Royal Astronomical Society* 67 (1981) 375–393.
[21] S.M. Day, *Bulletin of the Seismological Society of America* 72 (1982) 1881–1902.
[22] J. Virieux and R. Madariaga, *Bulletin of the Seismological Society of America* 72 (1982) 345–369.
[23] S. Das and K. Aki, *Journal of Geophysical Research* 82 (1977) 5658–5670.
[24] R. Madariaga, *Journal of Geophysical Research* 84 (1979) 2243–2250.
[25] R. Madariaga, *Proceedings of the International School of Physics* (1983) 1–44.
[26] R.J. Archuleta, *Journal of Geophysical Research* 89 (1983) 4559–4585.
[27] S. Das, "A Numerical Study of Rupture Propagating and Earthquake Source Mechanisms", Sc.D. thesis, Massachusetts Institute of Technology, Cambridge, MA (1976).
[28] C.C. Chao, *Journal of Applied Mechanics* 27 (1960) 559–567.
[29] P.G. Richards, *Bulletin of the Seismological Society of America* 69 (1979) 947–956.
[30] J.D. Achenbach, in *Mechanics Today* 1, Pergamon Press, New York (1973) 1–55.
[31] L.B. Freund, *Journal of Geophysical Research* 84 (1979) 2199–2209.
[32] R.J. Archuleta and S.M. Day, *Bulletin of the Seismological Society of America* 70 (1980) 671–689.
[33] L.B. Freund, *9th U.S. National Congress on Applied Mechanics* (1982).
[34] S. Das, *Bulletin of the Seismological Society of America* 72 (1982) 1911–1926.
[35] M.I. Huseini, D.B. Jovanovich, M.J. Randall and L.B. Freund, *Geophysical Journal of the Royal Astronomical Society* 43 (1975) 367–385.
[36] S. Das and B.V. Kostrov, *Journal of Geophysical Research* 88 (1983) 4277–4288.
[37] R.D. Mindlin, *Journal of Applied Mechanics* 16 (1949) 259–268.
[38] S. Das and B.V. Kostrov, *Geophysical Journal of the Royal Astronomical Society* (1985) in press.
[39] R. Burridge, in *Dynamic Waves in Civil Engineering*, Wiley (1971) 15–31.
[40] J.R. Rice, *Physics of the Earth's Interior* XXVIII (1980) 555–649.
[41] R. Madariaga, *SIAM-AMS Proceedings* 12 (1979) 59–77.
[42] R. Dmowska and J.R. Rice, in *Continuum Theories in Solid Earth Physics*, Elsevier Publishing Company (1985) 187–255.

Résumé

Un processus de glissement dans une faille souterraine peut être modélisé sous forme de fissure de cisaillement à 2 ou 3 dimensions en cours de propagation. Comme les méthodes analytiques ne peuvent pas être utilisées pour des fissures finies, on a considéré diverses approches numériques à ce problème parmi laquelle la méthode des intégrales numériques aux limites est décrite en détail. Le processus de croissance d'une fissure peut être déterminé par l'utilisation d'un critère de rupture. Comme les propriétés du matériau telles que la résistance à la rupture ou les propriétés de friction varient sur les surfaces des failles réelles, deux types de modèles complexes de glissement, le modèle "Barrière" et le modèle "Aspérité", sont discutés dans le contexte de la seismologie.

International Journal of Fracture 27 (1985) 277–298.

On the measurement of dynamic fracture toughnesses – a review of recent work

J.F. KALTHOFF

Fraunhofer-Institut für Werkstoffmechanik, Wöhlerstrasse 11, D-7800 Freiburg, Federal Republic of Germany

(Received September 1, 1984)

Abstract

The influence of dynamic effects on test procedures for measuring the crack arrest toughness and the impact fracture toughness is analyzed. It is shown that these dynamic effects can be significant: For cracks at arrest the stress condition is still dynamic and not static although the crack velocity has become zero. Dynamic effects become small only for small crack velocities or small crack jumps or for specially designed specimens. It is shown that the stress intensity factor history for cracks under impact loading cannot be adequately derived from instrumented impact data via static evaluation procedures. Only for large times to failure, resulting for small impact velocities and/or ductile material behavior do static approaches represent acceptable approximations. Reliable crack arrest and impact fracture toughness data can only be obtained by evaluation procedures which take the dynamic effects into account, e.g. by utilizing the reduced dynamic effects crack arrest test specimen or by applying the concept of impact response curves.

1. Introduction

When static fracture toughness data are measured two parameters are involved: length and force. The measurement of dynamic fracture toughness data requires the consideration of the additional parameter time. Influences of time on the material behavior are to be measured, but the parameter time can also have an influence on the measuring procedure. In general, dynamic measuring procedures have their roots in measuring procedures for the equivalent static values. Often, the static procedures are directly transferred to the dynamic cases without any changes or with only slight modifications. Very few fully dynamic procedures have been developed. In general it is not trivial to make a decision as to which measuring methodology is the appropriate one. Procedures based on quasistatic analyses in general are easy to apply, but they impose restrictions on the test parameters and limitations on the applicability range of the test. Fully dynamic procedures, on the other hand, give reliable results under arbitrary test conditions but they are necessarily more complicated than static ones.

This paper summarizes work carried out by the author and his colleagues to investigate the influences of dynamic effects on test procedures for measuring the crack arrest toughness and the impact fracture toughness. The behavior of cracks under impact loading is examined as well as that of propagating and subsequently arresting cracks. The dynamic influences are quantified by comparing the actual dynamic reactions at the crack tip with the equivalent static behavior. The conditions under which quasistatic measuring procedures can be applied with sufficient accuracy are discussed. For situations where quasistatic procedures are not applicable dynamic procedures are described which take inertial and kinetic effects into account.

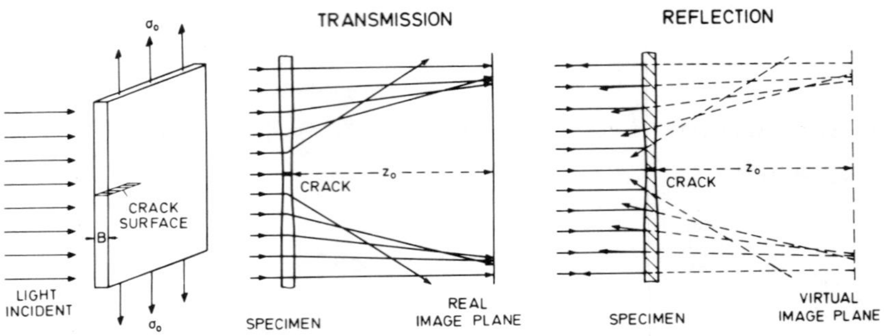

Figure 1. Physical principle of the shadow optical method of caustics.

The dynamic crack tip behavior is investigated by means of the shadow optical method of caustics. This method is briefly described first and fracture data are presented in the following sections.

2. The shadow optical method of caustics

The method of caustics is an optical tool for measuring stress intensifications. The technique was introduced by Manogg in 1964 [1,2] and has been developed further by other investigators [3–5]. The author and his coworkers extended and applied Manogg's technique for investigating dynamic fracture phenomena [6–9].

The physical principle of the method is illustrated in Fig. 1. A precracked specimen under load is illuminated by a parallel light beam. A cross-section through the specimen at the crack tip is shown in Fig. 1b for a transparent specimen, and in Fig. 1c for a

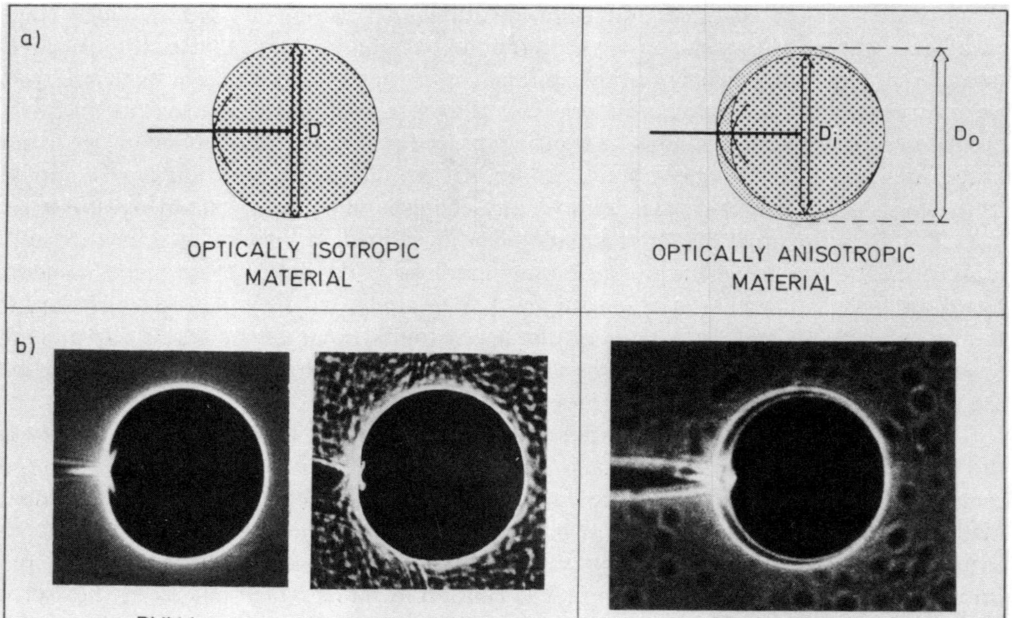

Figure 2. Mode I caustics, (a) calculated, (b) measured.

non-transparent steel specimen. Due to the stress concentration the physical conditions at the crack tip are changed. For transparent specimens both the thickness of the specimen and the refractive index of the material are reduced. Thus, the area surrounding the crack tip acts as a divergent lens and the light rays are deflected outwards. As a consequence, on a screen (image plane) at a distance z_0 behind the specimen a shadow area is observed which is surrounded by a luminous curve, the caustic (see Fig. 2). Figure 1c shows the situation for a non-transparent steel specimen with a mirrored front surface. Due to the surface deformations light rays near the crack tip are reflected towards the center line. An extension of the reflected light rays onto a virtual image plane at the distance z_0 behind the specimen results in a light configuration which is similar to the one obtained in transmission. Consequently a similar caustic is obtained. The mode I shadow pattern was calculated by Manogg [1] from the linearly elastic stress strain field around the crack tip. Figure 2 compares theoretical results with experimentally observed caustics which were photographed in transmission and in reflection with different materials. The single caustic curve obtained for isotropic materials splits up into a double caustic for optically anisotropic materials.

The size of the shadow pattern is related to the stress intensification at the crack tip. The quantitative correlation between the diameter D of the caustic and the stress intensity factor K_I is given by the relation

$$K_\text{I} = \frac{2\sqrt{2\pi} \cdot F(v)}{3 f_{\text{o,i}}^{5/2} \cdot c \cdot d_\text{eff} \cdot z_0} D^{5/2} \tag{1}$$

where

K_I = mode I stress intensity factor,
D = diameter of the caustic,
$f_{\text{o,i}}$ = numerical factor for outer/inner caustic,
c = photoelastic constant,
d_eff = effective thickness of the plate,
= d for transparent specimens,
= $d/2$ for reflecting, non-transparent specimens,
d = physical thickness of the plate,
z_0 = distance between specimen and image plane,
$F(v)$ = correction factor for non-zero crack velocities, $F(v) \lesssim 1$,
v = crack velocity.

Numerical values of the constants which appear in the K-evaluation formula are given in Table 1 for different materials. For further details of the shadow optical technique see [6] and [9].

Crack tip caustics are of a simple form and can be easily evaluated. The technique is, therefore, well suited for investigating complex phenomena, as e.g. in fracture dynamics. A Cranz Schardin 24 spark high speed camera is utilized in these investigations for photographing shadow patterns under dynamic loading conditions.

3. Arresting cracks

The usual procedure for measuring the crack arrest toughness K_{Ia} of a material is the following: In a wedge loaded specimen a rapidly propagating crack is initiated from a blunted initial notch at an initiation stress intensity factor $K_{\text{Iq}} > K_{\text{Ic}}$. Figure 3 shows a rectangular double cantilever beam specimen under longitudinal wedge loading in a schematic representation. The crack opening displacement remains constant during crack propagation due to the stiffness of the loading system. Thus the crack propagates into a decreasing stress intensity factor field. It arrests at the length a_a if the conditions for crack

Table 1. Constants for caustic evaluation

Material	Elastic constants		Shadow optical constants						Effective thickness
	Young's modulus MN/m^2	Poisson's ratio	for plane stress			for plane strain			
			c m^2/N	f_o	f_i	c m^2/N	f_o	f_i	d_{eff}
TRANSMISSION: $(z_0 < 0)$									
Optically anistropic:									
Araldite B	3660*	0.392*	-0.970×10^{-10}	3.31	3.05	-0.580×10^{-10}	3.41	2.99	d
CR-39	2580	0.443	-1.200×10^{-10}	3.25	3.10	-0.560×10^{-10}	3.33	3.04	d
Plate Glass	73900	0.231	-0.027×10^{-10}	3.43	2.98	-0.017×10^{-10}	3.62	2.97	d
Homalite 100	4820*	0.310*	-0.920×10^{-10}	3.23	3.11	-0.767×10^{-10}	3.24	3.10	d
Optically isotropic:									
PMMA	3240	0.350	-1.080×10^{-10}	3.17		-0.750×10^{-10}	3.17		d
REFLECTION: $(z_0 > 0)$									
All materials	E	ν	$2\nu/E$	3.17		–	–		$d/2$

* Dynamic value.

propagation are not fulfilled anymore. The stress intensity factor value at arrest represents the crack arrest toughness K_{Ia}.

3.1. Influences of dynamic effects

In order to investigate the influence of dynamic effects on the mechanical behavior of cracks at arrest, the stress condition at the tip of a propagating and subsequently arresting crack was analyzed [10–14]. Using the epoxy resin Araldite B, the actual dynamic stress intensity factors, K_I^{dyn}, were measured by means of the shadow optical method of caustics in transmission. These values are compared to the equivalent static stress intensity factors, K_I^{stat}, calculated from the measured crack opening displacement 2δ utilizing conventional stress intensity factor formulas from ASTM E 399.

A series of six shadow optical photographs is shown in Fig. 4. Quantitative data for cracks initiated at different K_{Iq}-values are presented in Fig. 5, where the dynamic, experimentally determined stress intensity factors, K_I^{dyn}, are shown as a function of crack length together with the corresponding static stress intensity factor curves, $K_I^{stat}(a)$. In addition, the measured crack velocities are given in the lower part of the diagram. The following characteristics of the crack arrest process can be deduced from these results: At the beginning of the crack propagation phase the dynamic stress intensity factor K_I^{dyn} is smaller than the corresponding static value K_I^{stat}. At the end of the propagation phase, in particular at the moment of arrest, the dynamic stress intensity factor K_I^{dyn} is larger than the corresponding static value K_I^{stat}. Only after arrest does the dynamic stress intensity factor K_I^{dyn} approach the static stress intensity factor at arrest, K_{Ia}^{stat}. Differences between the dynamic and the static stress intensity factor curves become smaller for cracks initiated at lower K_{Iq} values, i.e. for cracks propagating at lower velocities. The dynamic effects obviously decrease with decreasing velocity, as one might expect.

The behavior of the dynamic stress intensity factor, K_I^{dyn}, in the post-arrest phase is shown in Fig. 6 as a function of time. K_I^{dyn} oscillates around the value of the static stress intensity factor at arrest, K_{Ia}^{stat}. Only some time after arrest does the dynamic stress intensity factor approach the static value.

Wave propagation phenomena explain the observed behavior: Elastic waves are produced by the propagating crack, so that kinetic energy is radiated into the specimen and $K_I^{dyn} < K_I^{stat}$. After reflection at the finite boundaries of the specimen the waves interact with the crack again and contribute to the stress intensity factor, consequently $K_I^{dyn} > K_I^{stat}$.

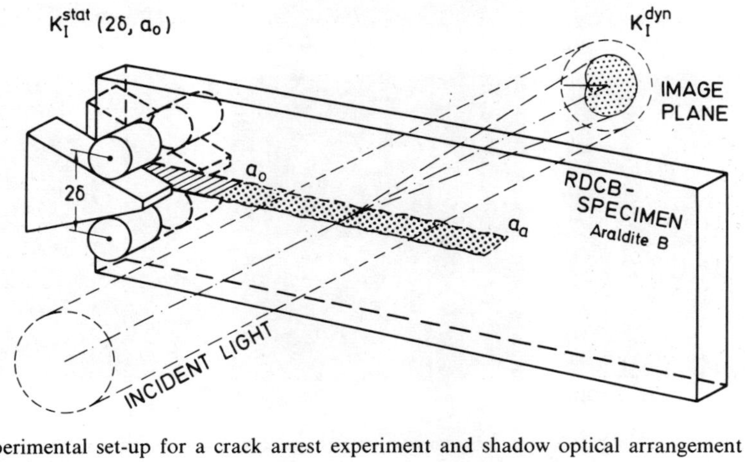

Figure 3. Experimental set-up for a crack arrest experiment and shadow optical arrangement in transmission (schematically).

An illustrative view of these processes is given in Fig. 7. A rapidly propagating crack (1000 m/s) in a high strength steel specimen was photographed in a shadow optical reflection arrangement [12]. In addition to the shadow spot at the crack tip the photograph shows the generation of waves at the tip of the propagating crack and the subsequent reflection at the boundaries of the specimen.

The findings are summarized in Fig. 8. The data show that the stress condition at arrest is not static but dynamic, although the crack velocity is zero at the moment of arrest. Crack arrest toughness values which are determined on the basis of a static evaluation procedure [15,16] cannot, therefore, in principle represent a true material property. Statically determined crack arrest toughness values K_{Ia}^{stat} are smaller than the true dynamically determined crack arrest toughness values K_{Ia}^{dyn}. Differences between statically and dynamically determined crack arrest toughness values depend on the shape and size of the specimen: They are very large for rectangular double cantilever beam specimens since the boundaries of the specimen are very near to the crack propagation path. Consequently the elastic waves can interfere with the crack after relatively small wave propagation paths [12]. The differences are considerably smaller in compact specimens, which are mostly used for crack arrest investigations, since the distances between the crack path and the boundaries are larger. But also for this type of specimen the differences can be quite remarkable for fast propagating cracks, i.e. for cracks arresting after large crack jump distances. This is demonstrated by the results shown in Fig. 9. The same trend is shown in Fig. 10 which summarizes the results of an ASTM cooperative test program on crack arrest toughness measurements with compact specimens (see data in [17]).

Safety predictions on the basis of K_{Ia}^{stat} data therefore can be erroneous. The determination of the true crack arrest toughness has to take the dynamic effects into account and requires a fully dynamic analysis, for example the application of the concept of recovered kinetic energy [18,19]. However, since the dynamic effects in large scale structures are in

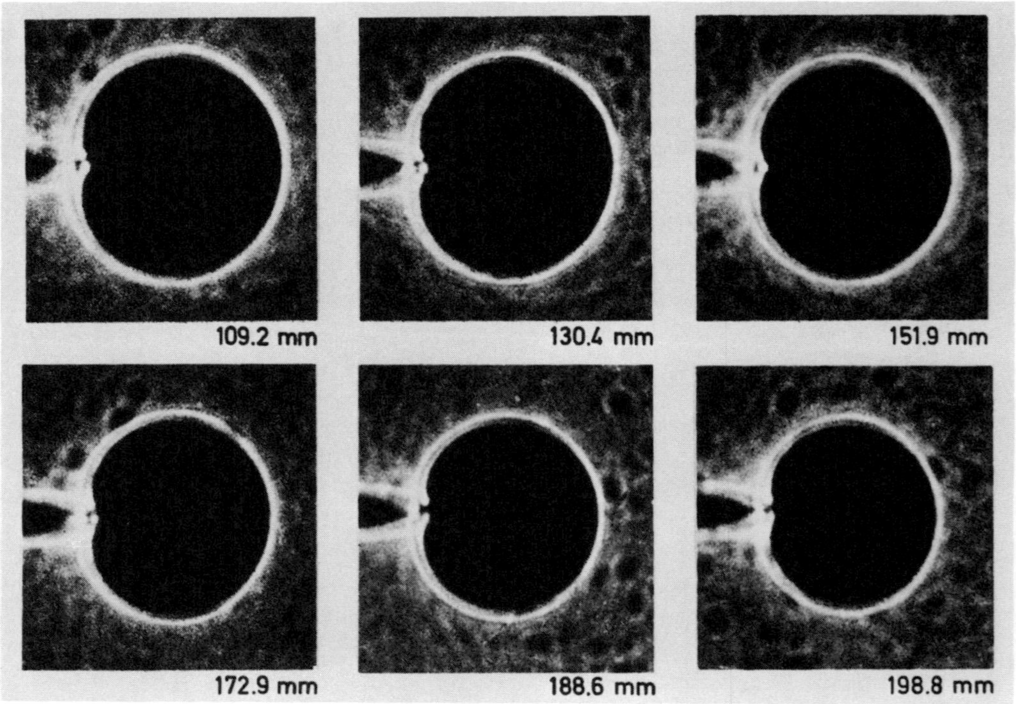

Figure 4. Caustics for a propagating and subsequently arresting crack (photographed in transmission with an Araldite B specimen).

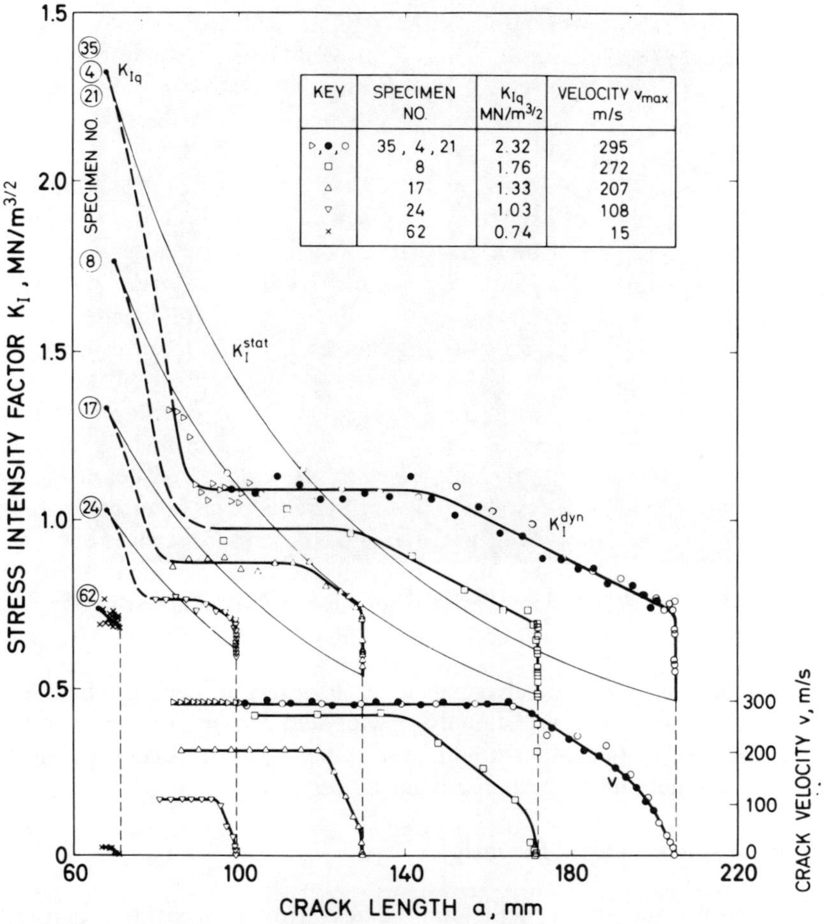

Figure 5. Stress intensity factors and crack velocity for propagating and subsequently arresting cracks.

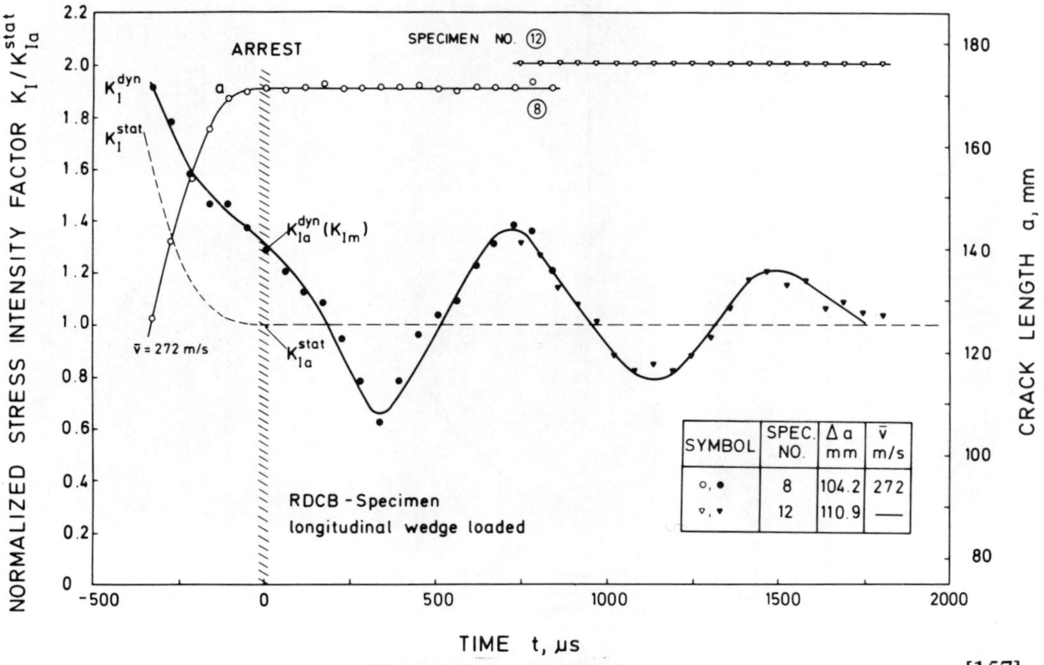

Figure 6. Post-arrest behavior.

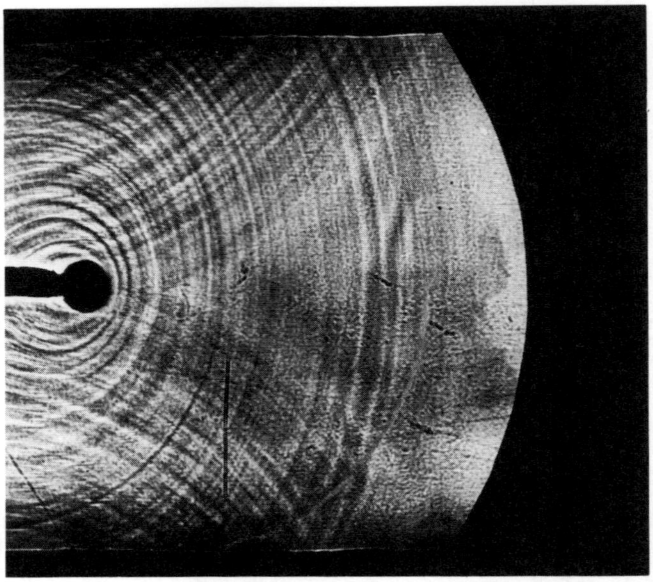

Figure 7. Shadow optical photograph of a fast propagating crack in steel.

general smaller than in the relatively small laboratory test specimens which are used for K_{Ia} determination, static crack arrest analyses will yield conservative crack arrest predictions [20]. On the basis of this understanding the static crack arrest concept can be applied by the practical engineer under certain circumstances.

3.2. *The reduced-dynamic-effects specimen*

In order to minimize the errors in the static crack arrest concept a special crack arrest specimen with reduced dynamic effects has been developed [21]. This reduced-dynamic-ef-

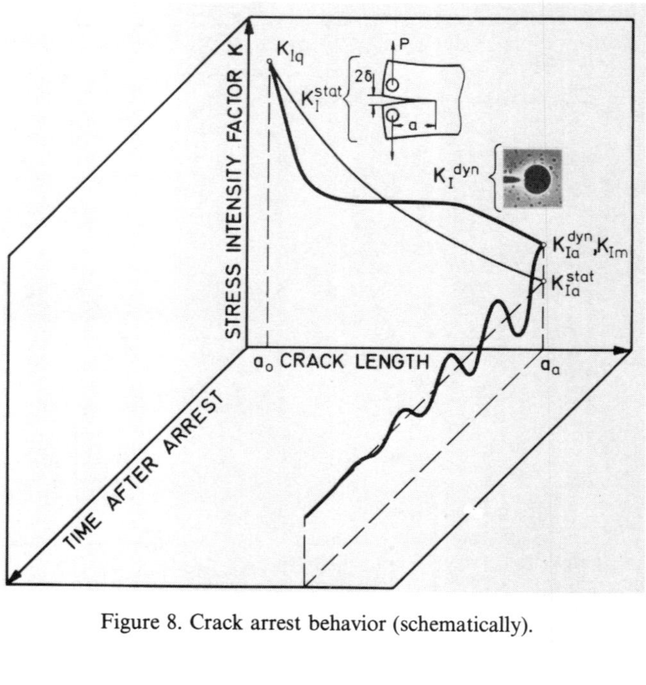

Figure 8. Crack arrest behavior (schematically).

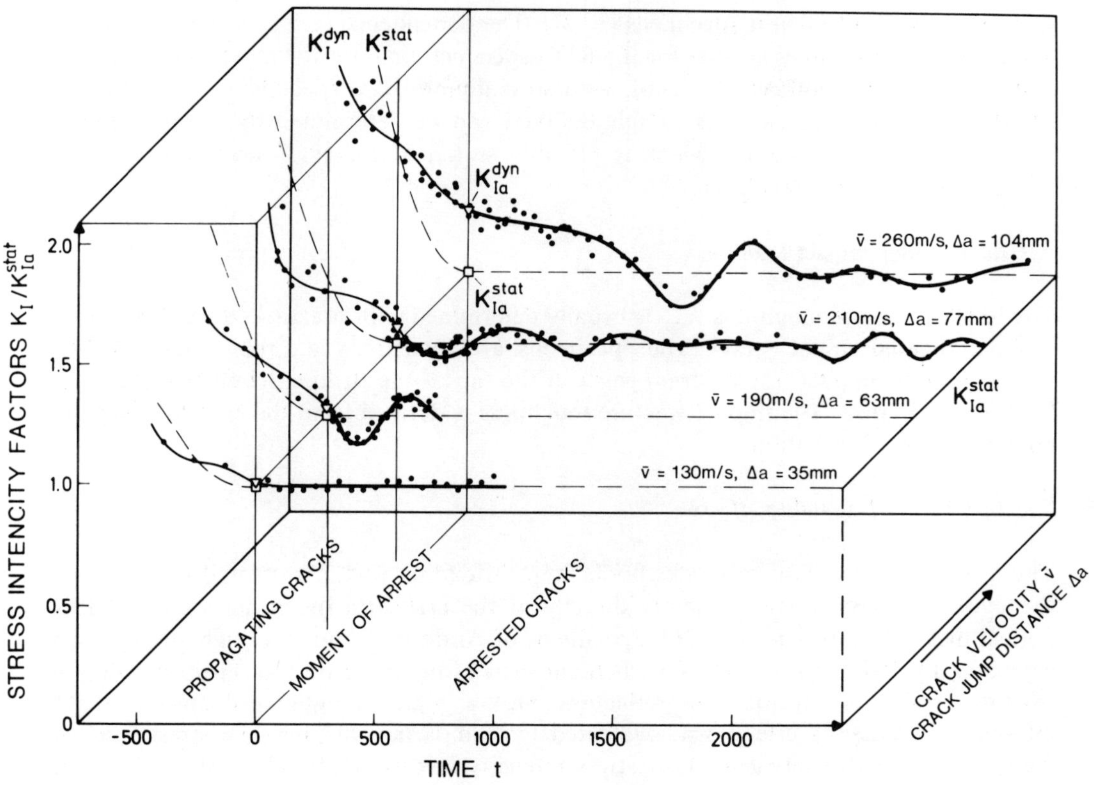

Figure 9. Crack arrest behavior observed with compact specimens.

fects-(RDE)-specimen is shown in Figs. 11 and 12. The edges and boundary of the specimen are shaped to reduce wave reflections and to defocus reflected waves. Damping material and additional weights are attached to the "wings" of the specimen to absorb kinetic energy and to increase the period of the eigenoscillation of the specimen, so that the recovery of kinetic energy is reduced. Statically determined crack arrest toughness values K_{Ia}^{stat} are shown in Fig. 13 for the RDE-specimen in comparison with data obtained

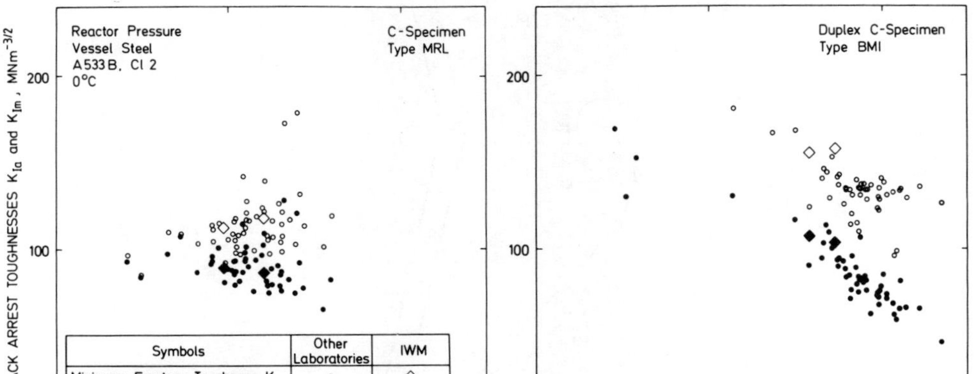

Figure 10. Crack arrest toughness data from ASTM Cooperative Test Program, $K_{Ia} \triangleq K_{Ia}^{stat}$ – determined according to a static analysis [15,16]; $K_{Im} \triangleq K_{Ia}^{dyn}$ – determined according to a fully dynamic analysis [18,19].

with other crack arrest test specimens [17,21]. The dependence on crack jump distance is about three to four times smaller for the RDE-specimen than for the most commonly used compact crack arrest specimen. Thus, with specially designed specimens, as for example the RDE-specimen, crack arrest toughness data can be determined by static evaluation procedures with an accuracy which is possibly sufficient for engineering purposes and which avoids over-conservatism.

4. Cracks under impact loading

The impact fracture toughness K_{Id} is usually determined with precracked bend specimens in instrumented impact tests. The specimens are loaded by a drop weight or by a pendulum type impact tester. Strain gages at the tup of the striking hammer measure the load during impact. The impact fracture toughness is derived from the critical load at the moment of crack initiation.

4.1. Influences of dynamic effects

The mechanical behavior of cracks under impact loading was investigated by measuring the dynamic stress intensity factors directly at the crack tip by means of the shadow optical method of caustics [22–26]. Specimens of Araldite B and of a high strength steel were investigated. Figure 14 gives a schematic view of the experimental set-up showing the shadow optical arrangement in reflection with a high strength steel specimen. The influence of dynamic effects was evaluated by comparing the dynamic stress intensity factors, K_I^{dyn}, with equivalent static stress intensity factors, K_I^{stat}. The static values were

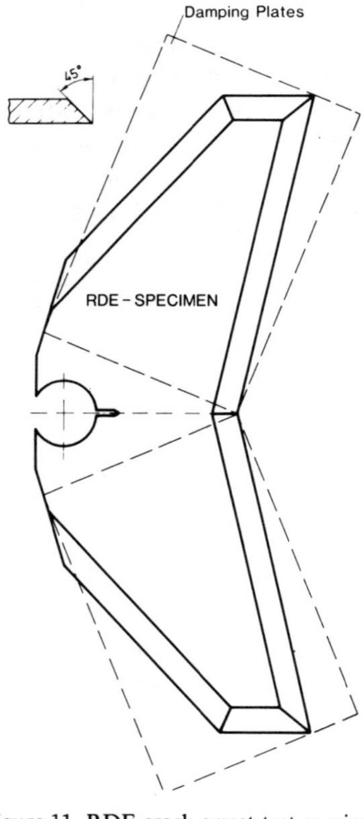

Figure 11. RDE-crack arrest test specimen.

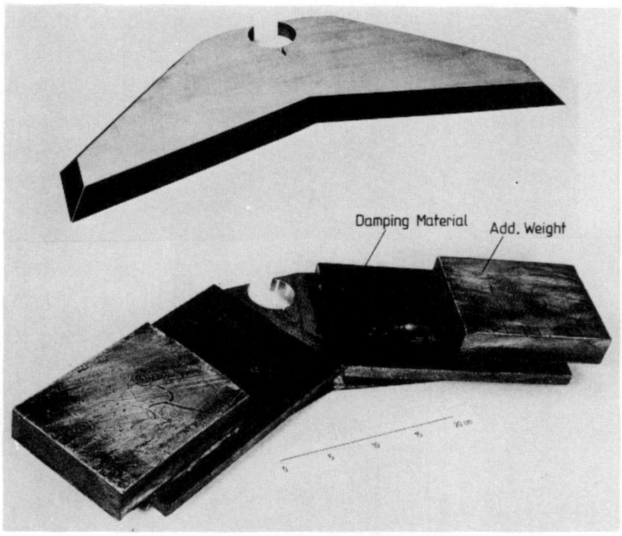

Figure 12. Photographs of RDE-crack arrest test specimen.

determined from the measured hammer load P_H utilizing conventional static stress intensity factor formulas from ASTM E 399.

A series of twelve shadow optical photographs of the central part of an impacted steel specimen is shown in Fig. 15. Quantitative data for the two materials considered are given in Fig. 16. The specimens were impacted at a velocity of 0.5 m/s by a hammer with a mass of 4.9 kg or 90 kg respectively. The dynamic stress intensity factors K_I^{dyn} (experimental points) and the corresponding static stress intensity factors K_I^{stat} (thin curves) are plotted as functions of time. The times are given in absolute units and also in relative units by normalization by the period τ of the eigenoscillation of the impacted specimen.

The K_I^{stat}-values show a strongly oscillating behavior, whereas the actual dynamic stress

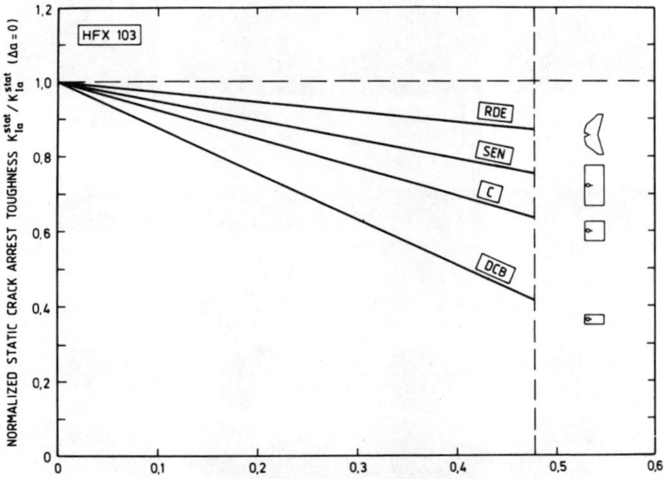

Figure 13. Dynamic effects for different crack arrest test specimens.

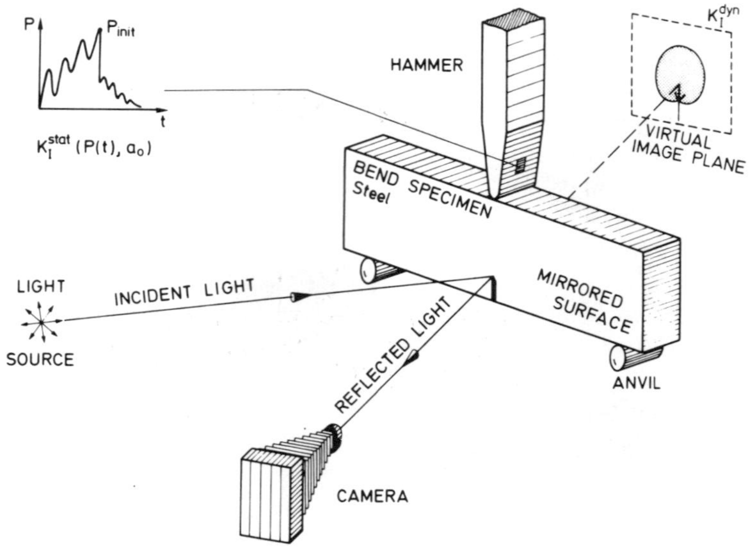

Figure 14. Experimental set-up for a drop weight experiment and shadow optical arrangement in reflection (schematically).

intensity factors K_I^{dyn} show a more steadily increasing tendency. In the small time range, $t < \tau$, these differences are very pronounced. The differences become smaller with increasing time, but even for times larger than 3τ the influence of dynamic effects obviously has

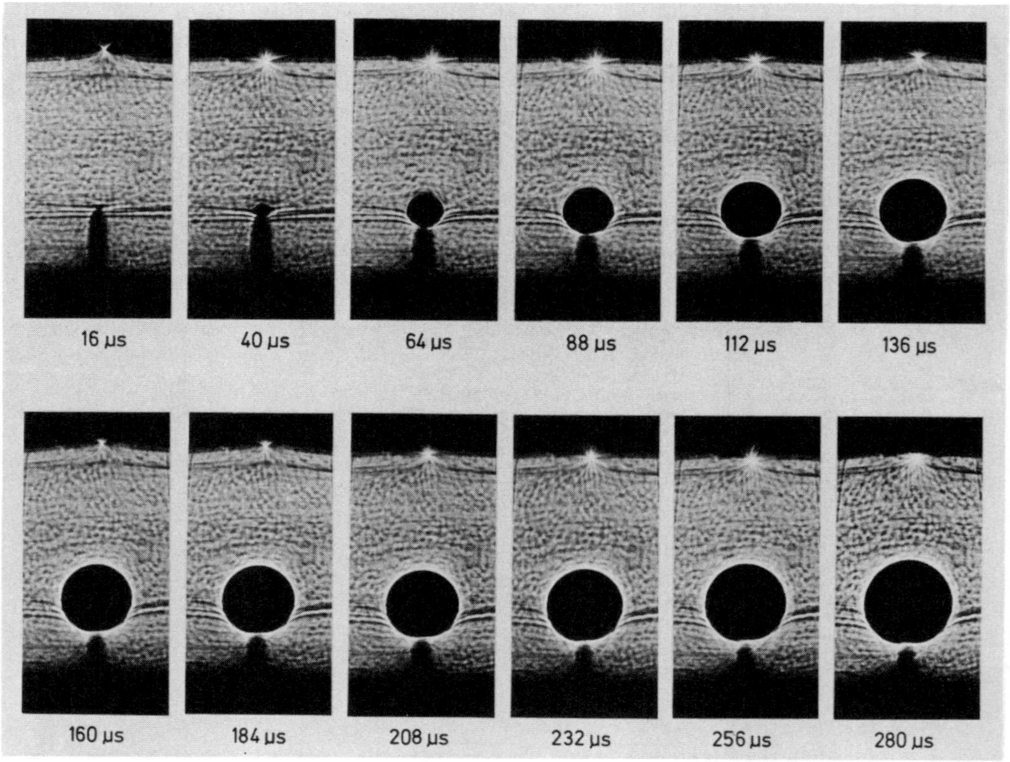

Figure 15. Shadow optical photographs of a precracked bend specimen during drop weight loading (photographed in reflection with a high strength steel specimen).

not vanished and there are still marked differences between K_I^{stat} and K_I^{dyn}. Due to a different contact stiffness the effects are larger with Araldite B specimens than with high strength steel specimens.

The behavior of specimens under impact loading was further investigated by also measuring the specimen reaction at the anvils [25]. Figure 17 compares the load measured at the striking hammer (a), the stress intensity factor measured at the crack tip (b), the load measured at the anvils (c), and the position of the specimen ends with regard to the anvils (d). The data were obtained with Araldite B specimens impacted at 1 m/s. The τ-value of the specimens utilized for these investigations is about 700 μs. A comparison of the four signals indicates that non-zero loads at the anvils are registered only after a rather long time of about 600 μs. This time is about three times larger than the time it would take the slowest wave, i.e. a transverse wave, to travel from the point of impact to the anvils. This unexpected behavior is explained by diagram (d) in Fig. 17. A loss of contact is observed between the specimen ends and the anvils. The loss of contact starts at about 200 μs. This time is in agreement with the above consideration of wave propagation times. For about the next 400 μs the specimen is completely free and only after this time, i.e. at a time of about 600 μs total do the specimen ends come into contact again with the anvils. In accordance with this observation loads are then recorded at the anvils. With different test conditions this loss of contact can occur later for a second time and loss of contact can also take place between the hammer and the specimen. These processes are illustrated in the schematic representation of Fig. 18. Since in these experiments the anvils were obviously of no influence during the early phase of the impact process, additional experiments were performed with unsupported specimens. The results are represented by the dashed curve and open data points in Fig. 17. In accordance with speculation, the early specimen reaction ($t < \tau$) is the same for both, the supported and the unsupported specimen.

The data demonstrate the strong influence of dynamic effects on the mechanical behavior of cracks under impact loading. A method [27] proposed to ASTM for measuring the dynamic fracture toughness K_{Id} in instrumented impact tests assumes that for times

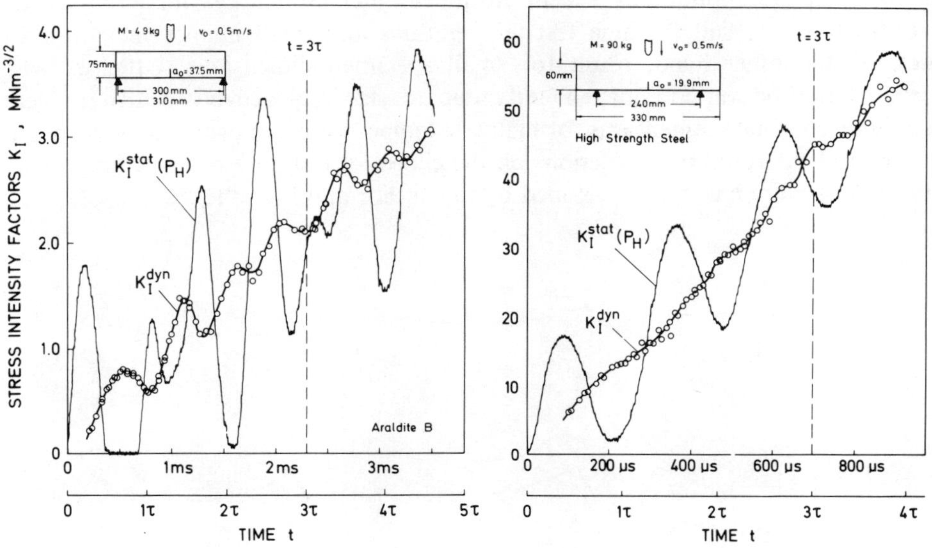

Figure 16. Stress intensity factors for cracks under drop weight loading: (a) Araldite B specimen, (b) high strength steel specimen.

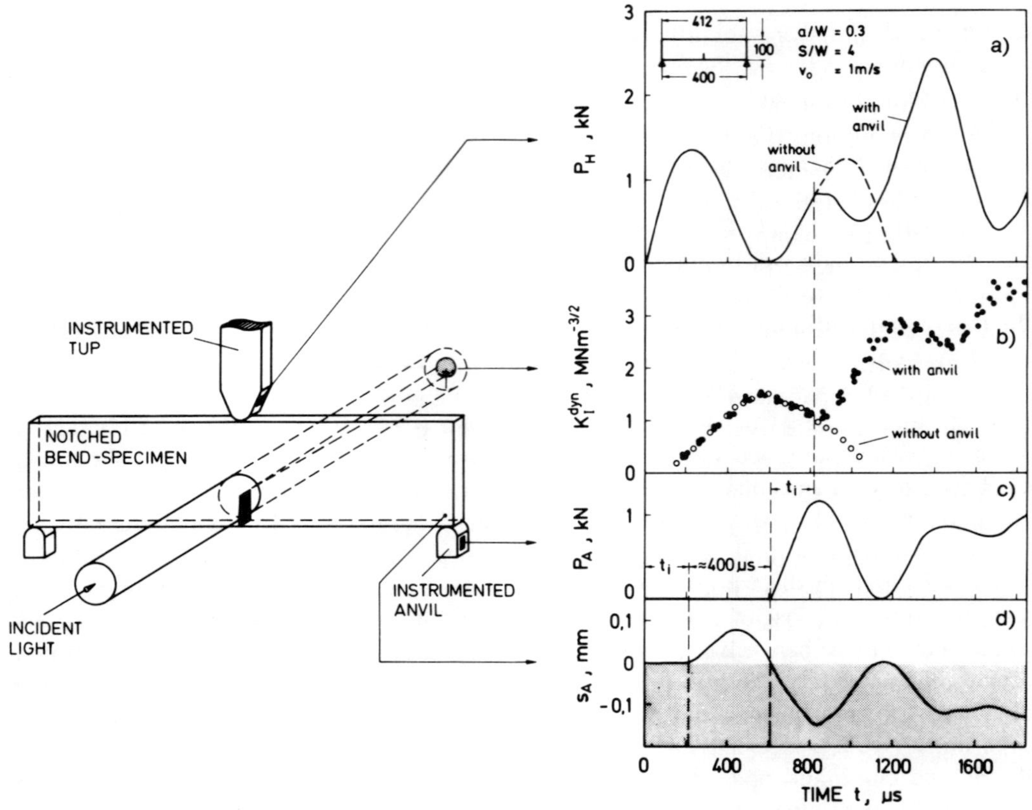

Figure 17. Mechanical response of a prenotched bend specimen to drop weight loading.

larger than 3τ K^{stat}-values would represent a good approximation of the actual dynamic stress intensity factor K_I^{dyn}. However, data from the presented experiments indicate that a static analysis is not at all adequate to describe the loading condition in the specimen under the proposed conditions except at much later times, i.e. for $t/\tau > 3$. Large times-to-fracture t_f result for high test temperatures and small impact velocities. Small τ-values, on the other hand, result for small specimen dimensions. Often t_f and τ, however, cannot be varied independently due to size requirements which in general demand large specimen dimensions for high test temperatures. In order to overcome these difficulties and the resulting restrictions on the choice of test parameters a fully dynamic measuring procedure has been developed by the author and his colleagues [28,29].

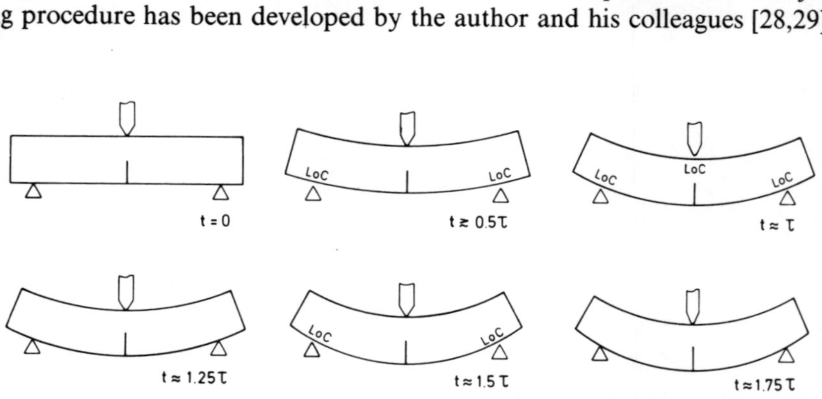

Figure 18. Loss of contact effects observed with prenotched bend specimens under drop weight loading.

4.2. The concept of impact response curves

The principle of the measuring procedure for this discussion is illustrated in Fig. 19. For fixed test conditions (i.e. specimen geometry, hammer mass, impact velocity, etc.) and a certain class of materials, e.g. steels, the dynamic stress intensity factor versus time relationship is determined in a pre-experiment such that $K_I^{dyn}(t)$ describes the response of the specimen to the impact process. This curve, called impact response curve, is controlled by the elastic properties of the system only. It there applies for all steels provided conditions for small scale yielding apply. In the real test-experiment a precracked specimen of the steel to be investigated is then tested under the same impact conditions. In this experiment only the time-to-fracture is measured. This time together with the preestablished impact response curve determines the impact fracture toughness value K_{Id}.

Impact response curves have been determined for various test geometries. Figure 20 gives data for large bend specimens of different types [26]. The data have been obtained with pre-notched specimens made from a high strengh steel utilizing the shadow optical method of caustics in reflection. Blunted notches instead of sharp cracks were utilized in these experiments in order to increase the load carrying capacity of the specimens before crack initiation. The specimens were impacted by a hammer of 90 kg or 260 kg mass at a velocity of 5 m/s. The hammer contact time determines $t = 0$. Specimens which were supported at the very ends of the specimen show a more oscillatory $K_I^{dyn}(t)$-curve. Specimens with a large length to support-span-ratio L/S = 55/40 (as with Charpy specimens) are characterized by a more steadily increasing curve (see also [25]). Figure 21 shows impact response curve data for Charpy specimens impacted at different velocities [29,30]. The specimens with notch length of 5 mm were tested in a 300 Joule pendulum type test machine (2 mm tup radius, machine compliance $C_M = 8.1 \times 10^{-9}$ m/N. The stress intensity factors were measured with a strain gage located at a small distance to the side of the crack tip. The strain gage signal was statically calibrated in terms of stress intensity factors.

The stress intensity factors K_I^{dyn} depend linearly on the impact velocity and are independent of the hammer mass provided the energy used for breaking the specimen is small compared to the total impact energy [31,32]. This condition is usually fulfilled with

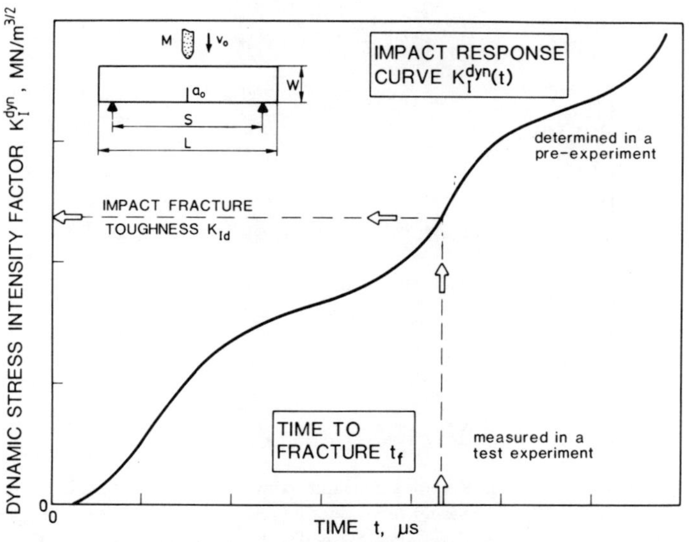

Figure 19. The concept of impact response curves for measuring the impact fracture toughness K_{Id}.

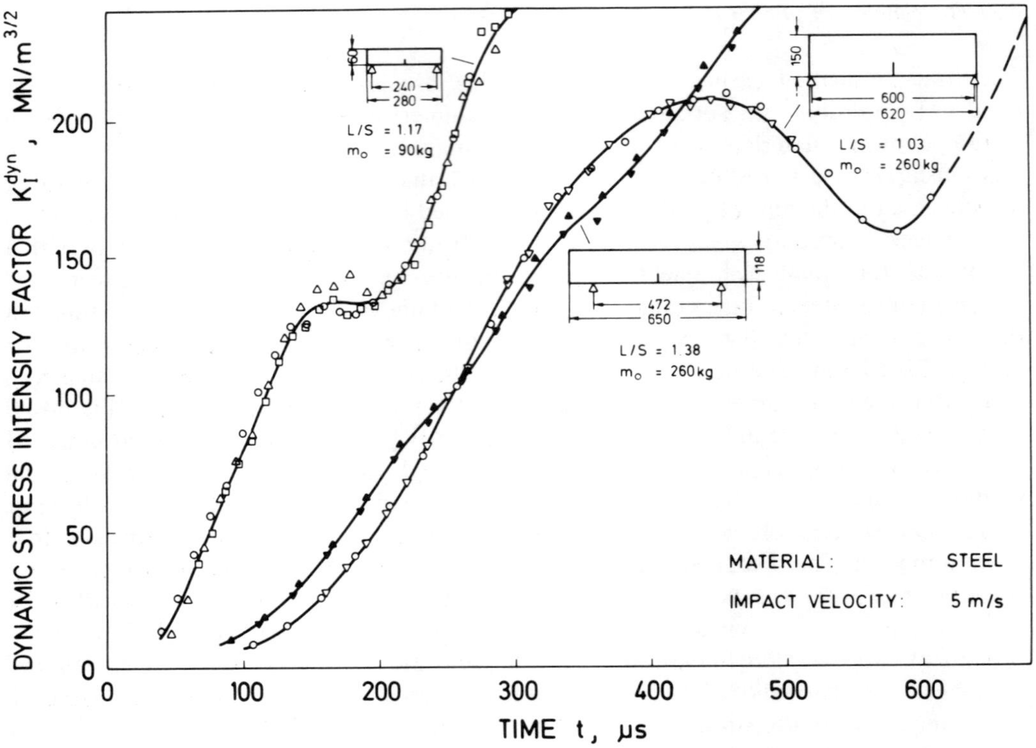

Figure 20. Impact response curves for precracked bend specimens.

conventional pendulum impact devices if the specimen breaks in the early time range. The impact response curves for arbitrary impact velocities can therefore be represented by a single relationship. For Charpy specimens this relation is given by (2) [29,30]. This equation is modified to account for small variations in crack lengths, $0.45 < a/W < 0.55$. The time is expressed in terms of the period of the apparent specimen oscillation.

$$K_I^{dyn} = R \cdot v \cdot \tau^* \cdot f(t/\tau^*) \quad \text{for} \quad 0 < t/\tau^* < 6 \quad \text{and} \quad 0.45 < a/W < 0.55 \tag{2}$$

where τ^* $= \tau/(1 + 1.4\,(a/W - 0.5))$
and τ $= 1.68\ (SBWCE)^{1/2} \cdot c_0$
with R $= 301\ GN/m^{5/2}$ [1]
v = impact velocity
τ = period of the apparent specimen oscillation (see [33])
a = crack length
W = specimen width
t = time
S = support span

[1] This value applies for stiff test devices with a machine compliance $C_M = 8.1 \times 10^{-9}$ m/N. If the actual compliance of the test device should differ from this value the resulting influence can be taken into account by multiplying the given value of R by the first order correction factor $1.276/(1 + 0.276\ C_M/8.1 \times 10^{-9}\ m/N)$. Procedures for determining the machine compliance of impact test devices are described in [35].

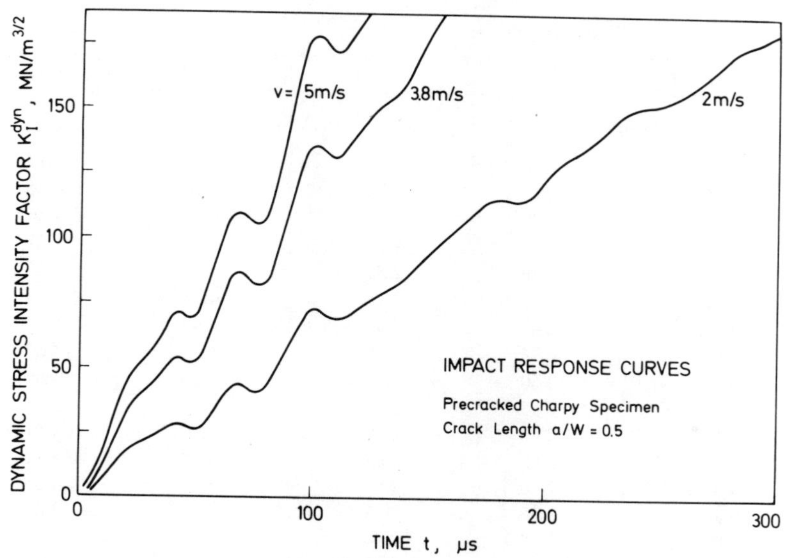

Figure 21. Impact response curves for Charpy specimens.

B = specimen thickness
C = specimen compliance (see [34])
E = Young's modulus
c_0 = sound wave speed (5000 m/s for steel)
$f(t/\tau^*)$ = function given by Table 2.

This approximate formula (see also [31,32]) describes the impact response curves for all practically relevant test conditions with an accuracy sufficient for engineering purposes.

With these impact response curves the impact fracture toughness of a structural steel is then determined by performing an impact experiment under the defined test conditions and measuring the time at which the crack becomes unstable. This time-to-fracture can be obtained from an uncalibrated strain gage near to the crack tip. The rapid decrease in the load signal is an indication of crack instability. Another procedure for measuring the time-to-fracture is based on the following phenomenon observed by the author and his colleagues. At the onset of rapid crack propagation a magnetic signal is generated by the accelerating crack tip if the specimen has been slightly magnetized before testing, e.g. with a permanent magnet [36]. This signal is picked up by a magnetic sensor (e.g. a coil) located at the crack tip near (but not in contact with) the specimen surface. [2] The obtained signal gives a clear indication of the time at the moment of fracture instability. This magnetic measuring procedure is inexpensive and advantageous for routine testing.

Impact fracture toughness data K_{Id} determined according to the described concept of impact response curves are shown for various steels and test temperatures in Fig. 22 a and b. A comparison with other fracture toughness data indicates that the measured values are reasonable [28,30]. This is not a trivial observation since the specimens in many cases failed at relatively small times-to-fracture, $t_f < \tau$, i.e. at times where the conventional procedure [27] would not apply.

The concept of impact response curves extends the conventional quasi-static evaluation procedure into the lower time-to-fracture range. It is an appropriate measuring tool for testing specimens which fail at relatively short times after impact, i.e. the procedure is

[2] Very good results are obtained with commercially available magnetic pickups as used in tape recorders.

Table 2. Time function for K_I^{dyn}-determination

t/τ^*	$f(t/\tau^*)$	t/τ^*	$f(t/\tau^*)$	t/τ^*	$f(t/\tau^*)$
0.00	0.0	2.00	2.34	4.00	4.12
0.05	0.0	2.05	2.30	4.05	4.20
0.10	0.04	2.10	2.25	4.10	4.23
0.15	0.11	2.15	2.24	4.15	4.26
0.20	0.19	2.20	2.25	4.20	4.28
0.25	0.28	2.25	2.29	4.25	4.31
0.30	0.37	2.30	2.33	4.30	4.37
0.35	0.47	2.35	2.38	4.35	4.44
0.40	0.55	2.40	2.43	4.40	4.52
0.45	0.61	2.45	2.48	4.45	4.60
0.50	0.65	2.50	2.52	4.50	4.67
0.55	0.69	2.55	2.56	4.55	4.74
0.60	0.73	2.60	2.59	4.60	4.81
0.65	0.77	2.65	2.62	4.65	4.84
0.70	0.82	2.70	2.67	4.70	4.85
0.75	0.88	2.75	2.73	4.75	4.85
0.80	0.91	2.80	2.80	4.80	4.86
0.85	0.93	2.85	2.91	4.85	4.87
0.90	0.92	2.90	3.00	4.90	4.89
0.95	0.90	2.95	3.06	4.95	4.93
1.00	0.90	3.00	3.13	5.00	4.98
1.05	0.93	3.05	3.19	5.05	5.03
1.10	1.01	3.10	3.23	5.10	5.08
1.15	1.11	3.15	3.29	5.15	5.13
1.20	1.27	3.20	3.35	5.20	5.17
1.25	1.38	3.25	3.42	5.25	5.22
1.30	1.45	3.30	3.49	5.30	5.27
1.35	1.46	3.35	3.56	5.35	5.34
1.40	1.43	3.40	3.63	5.40	5.41
1.45	1.38	3.45	3.68	5.45	5.47
1.50	1.37	3.50	3.71	5.50	5.53
1.55	1.38	3.55	3.73	5.55	5.61
1.60	1.44	3.60	3.70	5.60	5.68
1.65	1.55	3.65	3.67	5.65	5.75
1.70	1.67	3.70	3.65	5.70	5.79
1.75	1.80	3.75	3.67	5.75	5.82
1.80	1.96	3.80	3.73	5.80	5.84
1.85	2.12	3.85	3.82	5.85	5.86
1.90	2.25	3.90	3.92	5.90	5.88
1.95	2.34	3.95	4.02	5.95	5.92

particularly suitable for testing at high velocities and for testing brittle materials. The applicability range of the procedure is limited by the usual small scale yielding conditions, since the uniqueness of impact response curves is lost when large plastic deformations are present. Under the conditions of large plastic deformations, however, the resulting times-to-fracture in general are sufficiently long so that the quasi-static evaluation procedure can be successfully applied. Both procedures therefore have their specific ranges of applicability and complement each other.

4.3. Single point-bend test

In many of the experiments reported in Figs. 22 the specimens broke at times $t_f < \tau$. Consequently the anvils could not have influenced the specimen reaction. Since at high

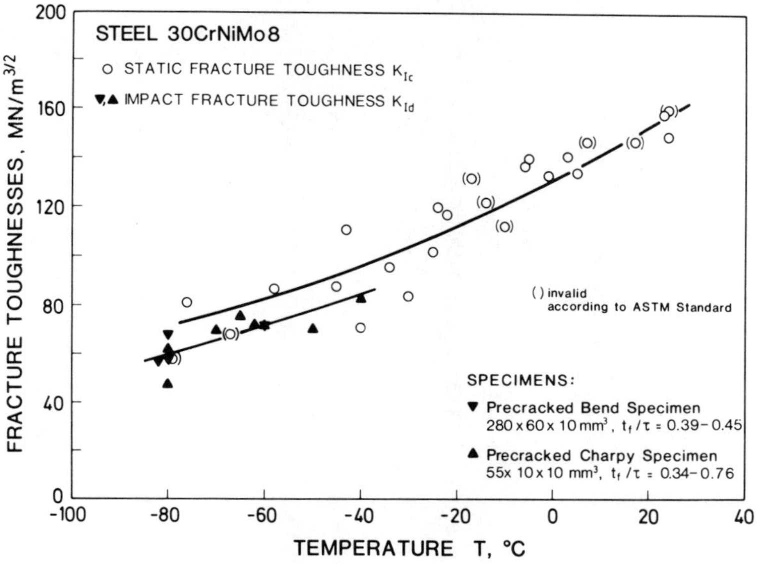

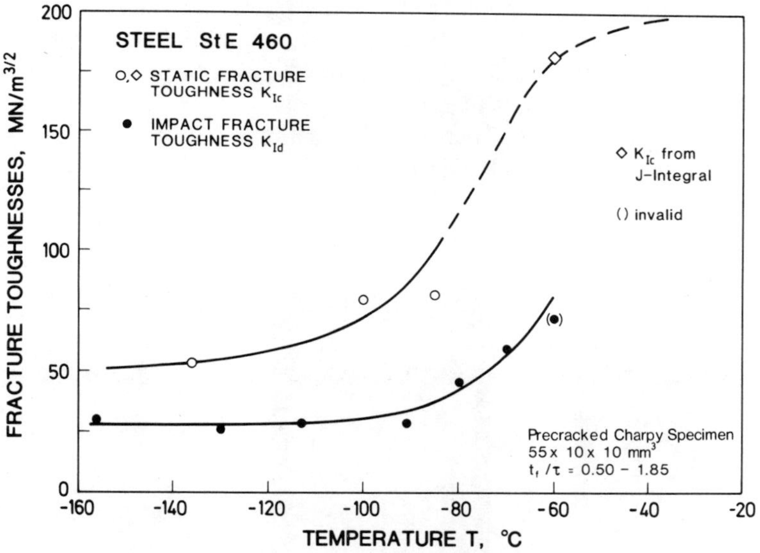

Figure 22. Impact fracture toughness data.

loading rates the specimens would fail even earlier it is natural that a high rate impact test ($>$ 5 m/s) for measuring K_{Id}-values can be carried out with unsupported specimens. This loading arrangement is called "single-point-bend" loading.

Typical results of "single-point-bend" experiments [26] are shown in Figs. 23 and 24. Steel specimens 270 mm long and 50 mm wide were utilized in these investigations. Impact loading was achieved by a projectile of 1.9 kg mass which was accelerated by a gas gun to a velocity of 17 m/s. The resulting impact response curve is shown in Fig. 23b. The stress intensity factor capacity of the loading arrangement is about 260 $MN/m^{3/2}$. Fracture toughness measurements have been performed under the same impact conditions with specimens made from a mild steel. Specimens 20 mm thick, with sharp initial notches, were tested at different temperatures. Complete failure of the specimens was

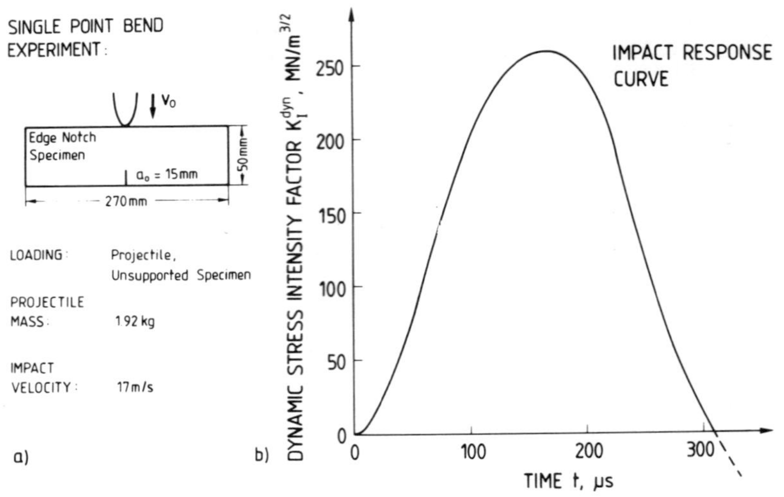

Figure 23. Single-point-bend technique, (a) test conditions, (b) impact response curve.

observed in all experiments. Figure 24 shows the fracture surfaces of the broken specimens, the times-to-fracture t_f, and the resulting dynamic fracture toughness values K_{Id}. These data demonstrate the feasibility of the single-point-bend technique for measuring impact fracture toughness data at high loading rates.

Similar experiments have been performed at a lower impact veloity of 5 m/s with a pendulum type impact tester [26]. The test device was slightly modified to load unsup-

Figure 24. Fracture behavior under single-point-bend loading.

ported specimens. At lower temperatures the specimens broke completeley in spite of the only modest impact velocity.

It is evident that a quasi-static evaluation procedure would not be able to describe these experiments. The fully dynamic method of impact response curves, however, can successfully be applied even under these highly dynamic test conditions.

5. Conclusion

The results presented here demonstrate that intuition based on a knowledge of statics can be misleading in dynamic situations. An appropriate consideration of the relevant time effects is needed for a successful treatment of dynamic fracture problems. Only on the basis of such fully dynamic investigations can one make decisions on the validity range of quasi-static measuring procedures or on the necessity for the application of fully dynamic test procedures.

References

[1] P. Manogg, "Anwendung der Schattenoptik zur Untersuchung des Zerreissvorgangs von Platten", Dissertation, Universität Freiburg, West Germany (1964).
[2] P. Manogg, in *Proceedings of the International Conference on the Physics of Non-Crystalline Solids,* Delft, Netherlands (1964) 481–490.
[3] P.S. Theocaris, *Journal of Applied Mechanics* 37 (1970) 409–415.
[4] P.S. Theocaris and N. Ioakimidis, *Journal of Applied Mechanics* 22 (1971) 876–890.
[5] A.J. Rosakis and L.B. Freund, *Journal of Engineering Materials and Technology* 104 (1982) 115–125.
[6] J. Beinert and J.F. Kalthoff, in *Mechanics of Fracture,* Vol. 7, ed. G.C. Sih, Martinus Nijhoff Publishers, The Hague, Boston, London (1981) 281–330.
[7] J.F. Kalthoff, in *Proceedings of the International Conference on Experimental Mechanics,* Society for Experimental Stress Analysis and Japan Society of Mechanical Engineers, Honolulu-Maui, Hawaii (1982) 1119–1126.
[8] J.F. Kalthoff, in *Workshop on Dynamic Fracture,* ed. W.G. Knauss, California Institute of Technology, Pasadena, California (1983) 11–35.
[9] J.F. Kalthoff, in *Handbook on Experimental Mechanics,* ed. A.S. Kobayashi, Prentice Hall, Englewood Cliffs, New Jersey, to appear in 1986.
[10] J.F. Kalthoff, S. Winkler and J. Beinert, *International Journal of Fracture* 12 (1976) RCR 317–319.
[11] J.F. Kalthoff, J. Beinert and S. Winkler, in ASTM STP 627 – *Fast Fracture and Crack Arrest*, American Society for Testing and Materials, Philadelphia (1977) 161–176.
[12] J.F. Kalthoff, J. Beinert, S. Winkler and W. Klemm, in ASTM STP 711 – *Crack Arrest Methodology and Applications,* American Society for Testing and Materials, Philadelphia (1980) 109–127.
[13] J.F. Kalthoff, *Fortschritt-Berichte der VDI-Zeitschriften,* Reihe 18, Nr. 4, VDI-Verlag, Düsseldorf (1978) 1–95.
[14] J.F. Kalthoff, J. Beinert and S. Winkler, "Influence of Dynamic Effects on Crack Arrest", Final Report prepared for Electric Power Research Institute, Palo Alto, California, under Contract No. RP 1022-1, IWM-Report W 4/80, Fraunhofer-Institut für Werkstoffmechanik, 78 Freiburg, West Germany (1980).
[15] P.B. Crosley and E.J. Ripling, *Nuclear Engineering and Design* 17 (1971) 32–45.
[16] P.B. Crosley and E.J. Ripling, in ASTM STP 627 - *Fast Fracture and Crack Arrest,* American Society for Testing and Materials, Philadelphia (1977) 372–391.
[17] J. Beinert, W. Klemm and J.F. Kalthoff, "Mechanik und Stoffverhalten bei der Arretierung von Rissen", IWM-Report V 16/83, prepared for European Community on Steel and Coal, under Contract No. 7210 Ke 107, Fraunhofer-Institut für Werkstoffmechanik, 78 Freiburg, West Germany (1983).
[18] G.T. Hahn, R.G. Hoagland, M.F. Kanninen and A.R. Rosenfield, in *Proceedings of an International Conference on Dynamic Crack Propagation,* ed. G.C. Sih, Lehigh University, Bethlehem (1972) 649–662.
[19] G.T. Hahn, P.C. Gehlen, R.G. Hoagland, M.F. Kanninen, C. Popelar, A.R. Rosenfield, et al., "Critical Experiments, Measurements and Analyses to Establish a Crack Arrest Methodology for Nuclear Pressure Vessel Steels", Reports BMI-1937, 1959, 1995, Battelle Columbus Laboratories, Columbus, Ohio (1975), (1976), (1978).
[20] J.F. Kalthoff, J. Beinert and S. Winkler, in *Proceedings 8. Sitzung, Arbeitskreis Bruchvorgänge im Deutschen Verband für Materialprüfung,* Köln, West Germany (1976) 138–144.

[21] J. Beinert and J.F. Kalthoff, in *Proceedings of International Conference on Application of Fracture Mechanics to Materials and Structures,* eds. G.C. Sih, S. Sommer, W. Dahl, Freiburg, West Germany (1983) 493–507.

[22] J.F. Kalthoff, S. Winkler and J. Beinert, *International Journal of Fracture* 13 (1979) RCR 528–531.

[23] J.F. Kalthoff, W. Böhme, S. Winkler and W. Klemm, in *Proceedings CSNI Specialist Meeting on Instrumented Precracked Charpy Testing,* EPRI, Palo Alto, California (1980) II 1–16.

[24] J.F. Kalthoff, W. Böhme and S. Winkler, in *Proceedings VIIth International Conference on Experimental Stress Analysis,* Society for Experimental Stress Analysis, Haifa, Israel (1982) 148–160.

[25] W. Böhme and J.F. Kalthoff, *International Journal of Fracture* 20 (1982) R139-143.

[26] J.F. Kalthoff, S. Winkler, W. Böhme and D.A. Shockey, in *Proceedings of International Conference on Dynamical Mechanical Properties and Fracture Dynamics of Engineering Materials,* Czechoslovak Academy of Sciences, Institute of Physical Metallurgy, Brno-Valtice, CSSR (1983) 1–16.

[27] ASTM E 24.03.03, "Proposed Standard Method of Test for Instrumented Impact Testing of Precracked Charpy Specimens of Metallic Materials", Draft 2d, American Society for Testing and Materials, Philadelphia (1981).

[28] J.F. Kalthoff, S. Winkler, W. Böhme and W. Klemm, in *Proceedings of 5th International Conference on Fracture,* Cannes, 1981, *Advances in Fracture Research,* ed. D. Francois et al., Pergamon Press, Oxford, New York (1980) 363–373.

[29] J.F. Kalthoff, "Concept of Impact Response Curves", to appear in *Metals Handbook,* Vol. 8, Mechanical Testing, American Society for Metals, Metals Park (1985).

[30] J.F. Kalthoff, S. Winkler and W. Böhme, "A Novel Procedure for Measuring the Impact Fracture Toughness K_{Id} with Precracked Charpy Specimens", to appear in Proceedings of International Conference on Mechanical and Physical Behavior of Materials under Dynamic Loading, Paris, France (1985).

[31] W. Böhme, "Eine einfache Methode zur Bestimmung der dynamischen Rissspitzenbeanspruchung bei schlagbelasteten Dreipunktbiegeproben", IWM-Report Z 2/84, Fraunhofer-Institut für Werkstoffmechanik, 78 Freiburg, West Germany (1984).

[32] W. Böhme and J.F. Kalthoff, "On the Quantification of Dynamic Effects in Impact Loading and the Practical Application for K_{Id}-Determination", to appear in Proceedings of International Conference on Mechanical and Physical Behavior of Materials under Dynamic Loading, Paris, France (1985).

[33] D.R. Ireland, in *Proceedings of International Conference on Dynamic Fracture Toughness*, London (1976) 47–62.

[34] R.J. Bucci, P.C. Paris, J.D. Landes and J.R. Rice, *Fracture Toughness, Proceedings, 1971 National Symposium on Fracture Mechanics*, Part II, STP 541, American Society for Testing and Materials, Philadelphia (1972) 40–69.

[35] D.R. Ireland, in ASTM STP 563 – *Instrumented Impact Testing,* American Society for Testing and Materials, Philadelphia (1974) 3–29.

[36] J.F. Kalthoff and S. Winkler, "Vorrichtung zur Erfassung des Rissstartes bei einer Bruchmechanik-Probe", Patentanmeldung P 33 34 570.8-52, Deutsches Patentamt München (1983).

Résumé

On a analysé l'influence des effets dynamiques sur les procédures d'essais pour mesurer la ténacité à l'arrêt d'une fissure et la ténacité à la rupture par impact. On montre que ces effets dynamiques peuvent être significatifs: pour des fissures qui s'arrêtent, les conditions d'essais sont toujours dynamiques et non statiques, bien que la vitesse de la fissure soit devenue égale à 0. Les effets dynamiques deviennent petits seulement pour de faibles vitesses de la fissure ou pour de petits resauts de la fissure ou encore dans le cas d'éprouvettes spécialement conçues. On montre que l'histoire du facteur d'intensité de contrainte pour des fissures soumises à une charge d'impact ne peut être convenablement déduite de données d'impact instrumenté évaluée selon une procédure statique. Ce n'est que pour de longues périodes de temps avant rupture, qui résultent de faible vitesse d'impact et/ou d'un comportement ductile du matériau que les approches statiques conduisent à des approximations satisfaisantes. Les données fiables de ténacité à l'arrêt d'une rupture ou à la rupture par choc ne peuvent être obtenues que par des procédures d'évaluation qui prennent en compte les effets dynamiques, par exemple en utilisant une éprouvette d'essai d'arrêt de rupture à effets dynamiques réduits ou en appliquant le concept des courbes réponses à l'impact.

International Journal of Fracture 27 (1985) 299–312.

Applications of dynamic fracture mechanics for the prediction of crack arrest in engineering structures

M.F. KANNINEN

Engineering and Materials Sciences Division, Southwest Research Institute, San Antonio, TX 78284, USA

(Received September 1, 1984)

Abstract

The compatibility of the static and dynamic approaches to crack propagation and arrest is investigated. For clarity, the term "kinetic" is introduced for analyses that focus on the entire crack initiation/propagation/arrest process – as opposed to static analyses which consider only the end points – as dynamic effects are not always significant in a crack run/arrest event. The importance of integrating experimental and computational work in this field is also discussed and a differentiation is given between the ways in which this can be done. An example computation on a thermal shock problem where dynamic effects are minimal, i.e., where the static and kinetic approaches are nearly the same, and a computation for a DCB specimen where significant differences occur between the two approaches are described. It is concluded that the static and kinetic approaches are entirely compatible as long as reflected stress waves do not reach the crack tip prior to crack arrest. But, when this is not the case, it is the kinetic approach that must be used. Similarly, when inelastic effects are important, only the kinetic approach can properly admit them.

1. Introduction

It might be thought that the term "dynamic fracture mechanics" only applies to those fracture problems in which inertia forces must be included in the equations of motion of the body. While such problems are certainly included, the subject is actually much broader. It encompasses all fracture mechanics problems where either the load or the crack size changes rapidly, regardless of whether or not inertia forces become significant. It follows that any time-dependent boundary value problem addressing rapid crack-growth initiation, propagation, and/or arrest lies within the domain of dynamic fracture mechanics. A phenomenological classification of problems in dynamic fracture mechanics might then include five specific focal points: the initiation of rapid crack propagation, the ensuing crack path, the crack speed, crack branching, and crack arrest.

Applications of the technology generally consider the latter four phenomena within a general theory of crack propagation. Thus, in practice, two kinds of dynamic fracture mechanics problems have received most attention: (1) bodies with stationary cracks that are subjected to a rapidly varying applied load, and (2) bodies under fixed or slowly varying loading that contain a rapidly moving crack. The basis for, and applications of, the second of these problem types is addressed in this paper. The emphasis in this paper is placed on the alternative methods for the prediction of crack arrest in elastodynamic conditions and the basis that these provide for addressing nonlinear material behavior. The treatment is drawn from the more comprehensive account provided by Kanninen and Popelar [1].

2. Background

Two mutually exclusive points of view for the arrest of a rapidly propagating crack have each been vigorously advanced. One stems from the suggestion, apparently made first by Irwin and Wells [2], that crack arrest can be considered as the reverse in time of crack initiation. This implies the existence of a material constant for crack arrest that is characterizable in terms of the *static* value of the crack driving force that corresponds to the crack length at the arrest point. Commonly, this parameter is the plane strain linear elastic fracture mechanics parameter K_{Ia}. Note that, since there is no necessity to consider the rapid crack propagation process preceding arrest – indeed, there is not even the possibility of so doing – this approach must be expressed in static terms.

The opposing point of view is based upon an equality between a measure of the crack driving force and the resistance of the material to crack extension. In linear elastic conditions a dynamically computed value of K drives the moving crack while a material and crack speed-dependent critical value of K, conventionally denoted by K_{ID}, is the resistance parameter. Crack arrest occurs when the equality between K and K_{ID} can no longer be maintained. This approach has previously been known as the dynamic view of crack arrest because of the dynamic effects that are included in the equations of motion of the cracked body. But, it is somewhat more accurate to refer to it as a *kinetic* approach because the focal point is on the crack growth process itself regardless of whether dynamic (inertia) forces are important or not.

Dynamic effects always exist in crack arrest. Nevertheless, statically-determined arrest values can be determined experimentally that often suffice for practical purposes. Specifically, the static approach requires a small crack jump length in the experimentation being conducted to determine the arrest property. The reason is that a dynamic analysis for crack propagation in an *infinite plane* provides the legitimization for the static view. Thus, the return of reflected stress waves to the crack tip prior to crack arrest can make the static approach invalid. As this condition is often not inhibiting in actual structures, the bulk of current state-of-the-art applications of dynamic fracture mechanics to rapid crack propagation and arrest are based upon the use of static crack arrest concepts. This section explores the basis for the use of the static and kinetic approaches to crack arrest and indicates where they are compatible and where they are not.

For the initiation of rapid crack propagation under linear elastic fracture mechanics (LEFM) conditions, one has the conventional LEFM relation for crack initiation, $K = K_{Ic}(T)$, where T denotes the temperature at the crack tip. In the static view of crack arrest, the termination of rapid crack propagation is expressed by

$$K = K_{Ia}(T). \tag{1}$$

In using (1), just as for initiation, K for the arrested crack is calculated as if the crack were stationary (quasi-static growth). And, like K_{Ic}, K_{Ia} is taken as a temperature-dependent material property that is independent of the component geometry. It can be re-emphasized that this point of view does not include any consideration of the unstable crack propagation process that links the initiation and arrest points.

The kinetic point of view gives direct consideration to crack propagation with crack arrest occurring only when continued propagation becomes impossible. Within the confines of elastodynamic behavior, unstable crack propagation occurs in such an approach under the condition that

$$K = K_{ID}(\dot{a}, T), \tag{2}$$

where K_{ID}, the dynamic fracture toughness, is a temperature-dependent function of the crack speed, $\dot{a}$, while K is the dynamically calculated value of the stress intensity factor.

Therefore, K must be determined as part of the solution of an initial value, (moving) boundary value problem. The K_{ID} function generally does not include an initiation value so that the equation $K = K_{Ic}$ is common to both approaches; i.e., $K_{ID}(0, T) \neq K_{Ic}(T)$. But, more significantly, K_{ID} does include an arrest value. In the kinetic approach, arrest occurs at the position and time for which the minimum value of K_{ID} exceeds K for all greater times.

As Barenblatt [3] has noted, crack propagation has a reasonably close analogy with a block pushed along a rough surface. The difference between static and sliding friction is similar to the difference between the initiation and the propagation fracture toughness. The block motion terminates (arrests) when it is not pushed firmly enough to overcome the sliding friction. Clearly then, within the framework of kinetic dynamic fracture mechanics, the counterpart of (1) is the inequality

$$K < K_{IA}(T), \tag{3}$$

where K_{IA} denotes the minimum value of K_{ID} with respect to crack speed.

Crosley and Ripling, using a static interpretation of the crack arrest process, provided the first extensive data on crack arrest [4]. While their original purpose was to obtain K_{Ic} data for rapidly applied loading (i.e., K_{Id} values), their work actually initiated quantitative treatments of crack arrest. Shortly thereafter, the work of Hahn and coworkers at Battelle's Columbus Laboratories initiated the kinetic approach to crack arrest [5]. While considerable data were developed, decisive evidence for one or the other points of view was lacking. This evidence was eventually forthcoming from the experiments on photoelastic materials performed by Kalthoff et al. [6] and by Kobayashi et al. [7]. These were instrumental in revolving the issue in favor of the kinetic point of view. An example is the key result of Kalthoff and coworkers that is shown in Fig. 1.

The experimental results shown in Fig. 1, obtained using the method of caustics combined with high-speed photography, enabled a direct measurement of the crack driving force for a fast moving crack to be obtained for the first time. It can be seen that there is a distinct difference between the actual arrest value and the static value that exists at a later time. Kalthoff's results also provided direct confirmation of the computational results obtained earlier by Kanninen [8]. These had indicated both the incompatibility of (1) and inequality (3) when reflected stress waves interact with the moving crack tip and the systematic decrease of K_{Ia} as the crack jump length increases. The latter trend, it

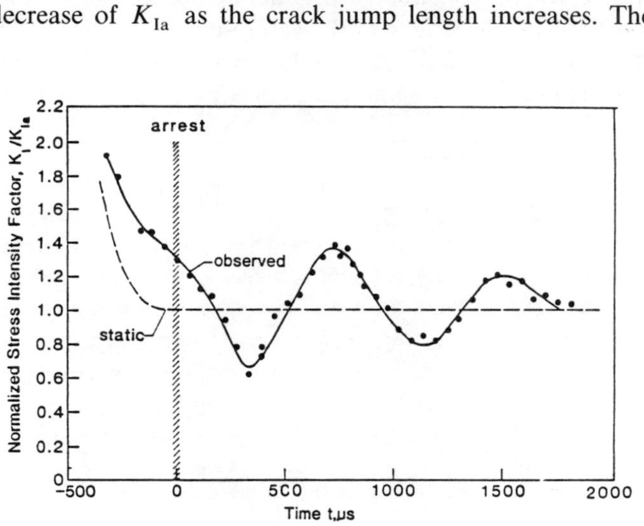

Figure 1. Comparison of static stress intensity factors and actual values during rapid crack propagation and following arrest in a DCB test specimen as determined by the method of caustics, results of Kalthoff et al. [6].

might be noted, is distinctly evident in the recently completed cooperative crack arrest testing program reported by Crosley et al. [9].

By performing experiments in different crack speed regimes, Kalthoff and his co-workers were also able to establish K_{ID} functions for polymers that were in general accord with those developed indirectly by Hahn et al. [5] for metals. Kalthoff's later development of the method of caustics in reflection allowed K_{ID} values to be directly obtained for metals also. Rosakis et al. [10], by extending this technique, developed data for 4340 steel that typified the general speed dependence of the K_{ID} function for all materials. Their result is shown in Fig. 2.

All of the KID measurements that have so far been made have the qualitative form of the data shown in Fig. 2. These can be correlated by the heuristic empirical relation suggested by Kanninen and Popelar [1]

$$K_{ID} = \frac{K_{IA}}{1 - \left(\frac{\dot{a}}{V_l}\right)^m}, \tag{4}$$

where K_{IA}, V_l and m are temperature-dependent material constants. These constants have a clear physical interpretation: V_l corresponds to a limiting crack speed, K_{IA} corresponds to the nearly constant value in the low crack speed regime, while m is a dimensionless shape factor. To illustrate, the 4340 steel data shown in Fig. 2 can be represented reasonably well by the simple values $V_l = 1100$ m/sec, $K_{IA} = 60$ MPa $m^{1/2}$ and $m = 2$ although, to be sure, other sets of constants would also suffice.

While undermining the conceptual basis for the static approach to crack arrest by establishing the kinetic approach, results of the type shown in Fig. 2 and representable by (2), also provide a pragmatic basis for the use of (1). It can be shown, provided a K_{Ia} value is obtained under the proper conditions, that it will be a good approximation to K_{IA}. Because this result is based on a solution for an infinite medium, as described in detail by Kanninen and Popelar [1], one must have crack jump lengths that are short enough so that reflected stress waves do not return to interfere with the moving crack tip. Then, a kinetic prediction of crack arrest will be indistinguishable from a static prediction. As a consequence, it now appears that the controversy over the use of static vs kinetic treatments of rapid crack propagation, which dominated dynamic fracture discussions during the decade of the 1970's, is now fairly well resolved.

As will be described more fully in the following section, kinetic treatments are necessary for certain geometries, e.g., where the crack propagation path parallels a nearby

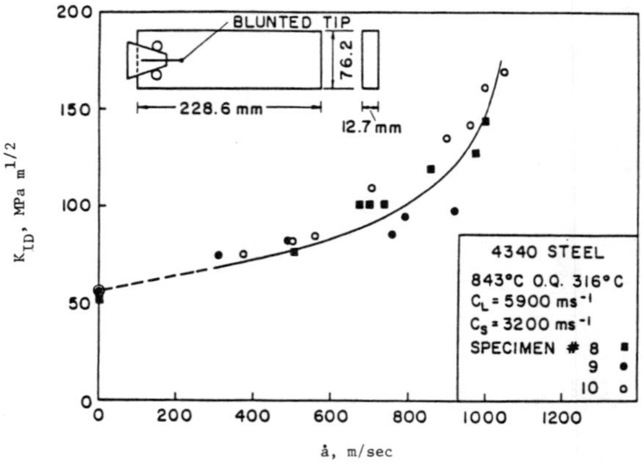

Figure 2. Dynamic fracture toughness values for 4340 steel determined by the method of caustics in reflection, results of Rosakis et al. [10].

free surface as in a DCB or CT specimen; yet the simpler static point of view suffices for engineering structures in many types of applications. Replacing this controversy at the forefront of research interest is one involving the appropriate nonlinear characterization for propagating cracks in the dynamic-viscoplastic conditions that would correspond to crack propagation and arrest for ductile upper shelf material behavior. It might be recognized that, just as in the evaluation of dynamic effects, the kinetic approach is uniquely able to incorporate time dependent inelastic material behavior. It therefore provides the proper vehicle for assessing these effects.

3. Dynamic crack propagation arrest applications

To illustrate the application of dynamic fracture mechanics concepts to crack propagation and arrest – and, incidentally, to assess the importance of dynamic effects on crack arrest – three applications are considered in the following. The first is a DCB test specimen, the second a pressurized gas transmission pipeline, while the third is for a thermal shock event in a nuclear power plant pressure vessel. Figure 3 illustrates the two different types of computations that can be performed. The analyses described in the following are of the application-phase type.

Analysis of a DCB specimen

The double cantilever beam (DCB) test specimen has been used extensively throughout the history of crack arrest testing. But, only recently has it become widely known that the DCB specimen, because of the proximity to the crack tip of the two free surfaces that parallel the crack plane, is the most dynamic of all of the configurations now in use. While this should constrain the use of a static analysis in conjunction with DCB specimen

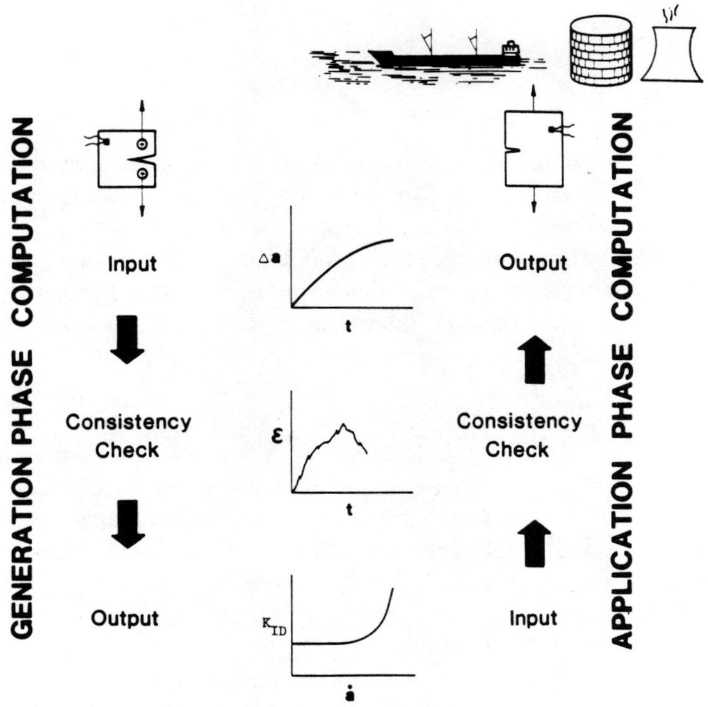

Figure 3. Integrated experimental/computational analysis approaches for dynamic crack propagation and arrest.

testing, it presents no difficulty for a kinetic analysis procedure. In fact, DCB specimen testing offers an excellent opportunity to study dynamic effects on crack arrest owing to their pervasive nature in this specimen.

A convenient arrangement for testing tough materials is through the use of a duplex specimen. In this arrangement, a fast running crack is initiated from a blunted crack in a low toughness starter section that is welded to the material to be tested. In this way the speed of the crack that enters the test section can be varied arbitrarily. To analyze results obtained in this way, the beam-like character of the DCB specimen was exploited by a number of investigators who developed expressions for the energy release rate using simple cantilever beam relations. Kanninen [11] extended this idea by modelling the uncracked portion of the specimen as a beam on an elastic foundation. As a check, an expression for the specimen compliance for the entire range of possible crack lengths could be obtained. In terms of an applied load P, the stress intensity factor deduced for this model is, with "a" as the cracklength

$$K = 2\sqrt{3}\,\frac{Pa}{Bh^{3/2}}\left\{\frac{\sinh^2\lambda c + \sin^2\lambda c}{\sinh^2\lambda c + \sin^2\lambda c} + \frac{1}{\lambda a}\,\frac{\sinh\lambda c\cosh\lambda c - \sin\lambda c\cos\lambda c}{\sin^2\lambda c - \sin^2\lambda c}\right\}, \tag{5}$$

where $2h$ and B, respectively, are the specimen height and thickness, c is the uncracked length portion of the specimen, and $\lambda = (6)^{1/4}/h$. Equation (5) was found to be in excellent agreement with the numerical results of Srawley and Gross [12]. More recently, Fichter [13] has used a Fourier transform and the Wiener-Hopf technique to determine the stress intensity factor for plane stress conditions. Because he neglected the free end, a comparison can be made only with the special case where $c \gg h$. For comparison, the three results can be written for this case as

$$\frac{Kh^{1/2}}{2\sqrt{3}\,P} = \begin{cases} \dfrac{a}{h} + 0.69, & \text{ref. [12]} \\ \dfrac{a}{h} + 0.6728 + 0.0377\dfrac{h^2}{a^2}, & \text{ref. [13]}. \\ \dfrac{a}{h} + 0.64, & \text{ref. [11]} \end{cases} \tag{6}$$

It can be seen that for practical purposes these results are essentially the same. Equation (5), which provides the only closed form expression for the complete specimen, indicates that free end effects cannot be neglected unless $c > 2h$. More important, it provides the only vehicle for considering dynamic crack propagation, albeit in a one-dimensional sense.

For dynamic conditions, the energy balance condition with kinetic energy included was taken as the crack growth criterion. This can be written, per unit thickness, as

$$G = \frac{\mathrm{d}W}{\mathrm{d}a} - \frac{\mathrm{d}U}{\mathrm{d}a} - \frac{\mathrm{d}T}{\mathrm{d}a}, \tag{7}$$

where, as above, W and U denote the external work done on the component and its elastic energy, per unit thickness, respectively, while T denotes the kinetic energy in the component per unit thickness. Crack growth is then supposed to proceed according to a relation $G(t) = G_c(\dot{a})$ with crack arrest occurring at a time when this criterion is no longer satisfied. For elastodynamic problems, the dynamic stress intensity factor is obtained from a dynamic generalization of the conventional J integral via $K = [EJ/(1-\nu^2)]^{1/2}$ which neglects a small crack speed correction.

The modern approach to the analysis of structural components rests on the finite element method. The approach is nevertheless much the same. In particular, in order to perform an application-phase computation, fracture toughness values must be known. As an example, typical values are

$$K_{ID} = \begin{cases} 65 + 0.044\, a, & \text{starter section material} \\ 154 + 0.077\, a^2, & \text{test section material} \end{cases}, \tag{8}$$

which correspond roughly to the use of 4340 steel in the low toughness starter section and A533B steel as the test material. In (8) the units of $\dot{a}$ are m/sec while the resulting K_{ID} values are in MPa $m^{1/2}$. A computational result obtained with these values by J. Ahmad is shown in Fig. 4.

For the specific calculation shown in Fig. 4, K_0, the stress intensity factor at the initiation of rapid crack growth, was selected such that the static K value at the point of entry into the test section would be just equal to the K_{IA} value of the test material; a fixed load point displacement was used. Thus, assuming $K_{Ia} = K_{IA}$, the static approach would predict arrest at the interface between the starter section and the test section. However, as shown by the dashed line in Fig. 4, the kinetic computation predicts that the crack would penetrate a considerable distance into the test material. Specifically, while the static approach predicts a total crack jump from 90 mm (the initial crack length) to only 123 mm (the interface), the kinetic approach predicts additional propagation in the test material to a total crack length of 230 mm. The difference is quite obviously very large.

Suppose for illustrative purposes that the computed dynamic result shown in Fig. 4 is an actual test record. Then, if a generation-phase analysis were to be performed, a kinetic analysis could be used to help establish (8). It is important to recognize that this could be done whether or not the crack actually arrested. The approach would readily handle the stress wave interaction with the propagating crack tip that is engendered in such an experiment. Recognition of this fact obviates the need to devise specimen types that "trap" stress waves in order to legitimize static analyses.

If a static interpretation was made of the dynamic result in Fig. 4 as if it were an experiment, it would lead to a value of $K_{Ia} = 45$ MPa $m^{1/2}$; i.e., the static K value at a crack length of 230 mm. This is obviously a conservative value, but one that considerably undervalues the true toughness property of the material as given by $K_{IA} = 154$ MPa $m^{1/2}$.

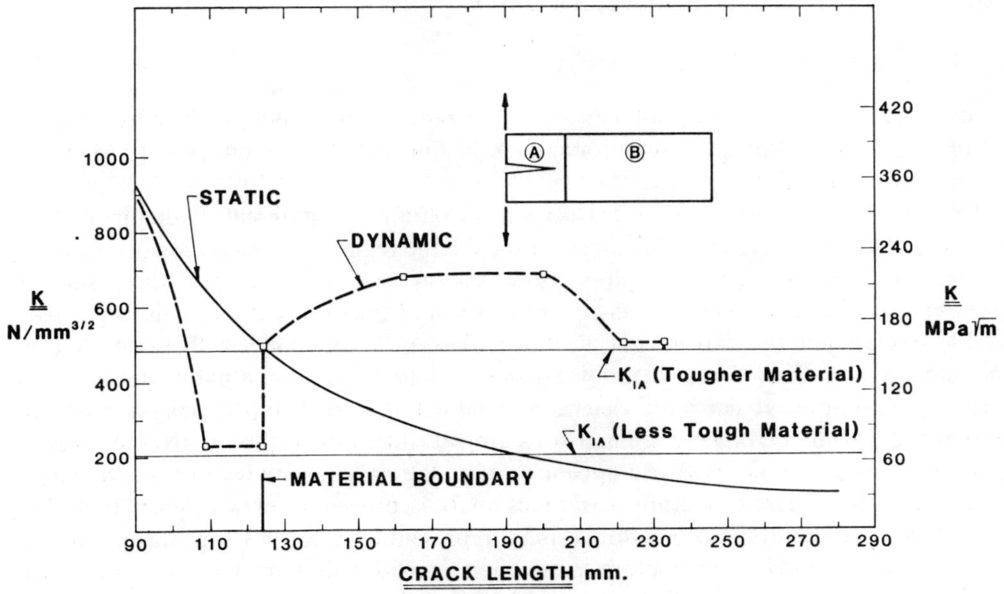

Figure 4. Comparison of static and dynamic computations for a duplex DCB specimen.

Taken together with the foregoing, this observation reinforces – in an exaggered fashion – the basic conclusion that can be drawn in regard to dynamic effects: It is generally conservative to draw a K_{Ia} value from a static interpretation of a laboratory crack run/arrest experiment, but non-conservative to apply a static analysis in a structural integrity assessment.

Finally, as a rationale for allowing the specimen friction effects that are known to occur in split-pin loaded CTA specimen testing, it has been suggested that these act to "damp" the oscillations in the dynamic stress intensity factor at arrest; c.f., Fig. 1. The static and the dynamic K values will then be nearly identical, thus eliminating the need for dynamic generation-phase analyses. However, what is overlooked in such an argument is that the static value achieved when friction forces are present might well be different from that when they are absent as they are in the structural component. The results that have so far been obtained suggest that this may be so and, moreover, that the K_{Ia} values inferred from tests involving friction are artifically inflated. This reinforces the need to conduct experiments that can be interpreted via well-posed initial-boundary value problems.

Crack propagation in a pressurized pipeline

As illustrated by the full-scale tests conducted by Maxey et al. [14], crack propagation in pressurized pipelines generally occurs at an essentally constant speed. When arrest takes place, it does so in a fairly abrupt manner. Typically, the ductile (or shear) crack speeds observed in full-scale tests range from 100 to 300 m/sec, brittle crack propagation speeds from 600 to 1000 m/sec. Note that for typical gas transmission pipe dimensions (e.g., $R/h = 40$) these are considerably smaller than the Rayleigh wave speed, C_R.

Kanninen [15] provided a limiting speed prediction for rapid crack propagation in a gas transmission pipeline using a one-dimensional model. This approach was based on an elastic-plastic extension of Yoffe's branching criterion and a model based upon the analogy between the deformation of a circular cylindrical shell under axisymmetric loading and the deflection of beam on an elastic foundation. Under the assumptions that, (1) the crack speed exceeds the gas decompression speed, so that the pressure behind the propagating crack tip is zero and (2) the presence of the crack uniformly reduces the shell stiffness, the limiting crack speed was found to be

$$V_l = \frac{3}{4} C_0 \left(\frac{h}{R} \right)^{1/2} \qquad C_0 = \text{bar velocity}, \tag{9}$$

where h and R are the pipe wall thickness and radius, respectively. Comparisons with measured speeds of cleavage crack propagation in full scale tests conducted by Maxey et al. [14] agree well with this result.

Most experimental work has been focused on obtaining empirical guidelines for the toughness necessary to insure crack arrest in the ductile regime. Empirical results based on a minimum value of the Charpy upper shelf energy are available from the results of Maxey et al. and others. While a decisive comparison of the different relations might seem possible experimentally, this is not the case. The difficulty lies in the fact that no experiment has yet been able to determine a specific value of an arrest parameter for given operating conditions. It can only determine whether the crack has propagated or not. Consequently, while qualitative comparisons are possible, direct quantitative verification is not. A theoretical analysis would appear to offer the only way to resolve this dilemma. Accordingly, the dynamic fracture mechanics analysis procedure of Kanninen et al. [16] for pipelines that results in a one-dimensional representation for crack propagation is of interest. Under steady state conditions, explicit consideration was given both to the principal features of the gas filled pipe problem and the important elements of dynamic crack propagation.

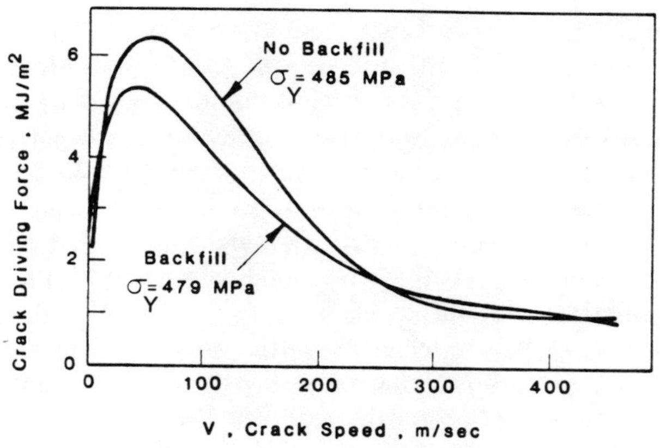

Figure 5. Calculated crack driving force for crack propagation in a pressurized pipeline ($R = 0.457$ m, h = 8.4 mm, $p = 8$ MPa), results of Kanninen et al. [16].

Starting from the equations for a circular cylindrical shell, four key assumptions were introduced by Kanninen et al. [16] to make the analysis tractable. These are that (1) radial deformations predominate, (2) circumferential variations in pressure can be neglected, (3) the crack-opening displacement is equal to the circumferentially integrated radial displacement $\bar{w}$ at any cross section in the cracked region, and (4) a plastic yield hinge is developed behing the crack tip. Further simplification was introduced by specializing to steady-state conditions. This was accomplished with the coordinate transformation $\xi = x - \dot{a}t$, where x is the axial coordinate, t is time, and $\dot{a}$ is the crack speed. This led to an ordinary differential equation that involved the pressure exerted on the pipe walls. Hence, a fluid mechanics treatment was also required. By assuming a predominately axial flow problem and accounting for the change of pipe cross-sectional area plus gas leakage behind the crack tip, an equation governing the pressure distribution $p = p(\xi, \dot{a})$ was obtained to decouple the problem.

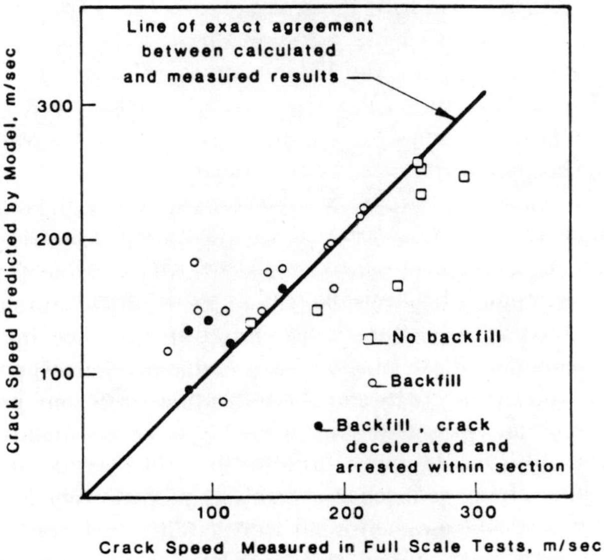

Figure 6. Comparison of predicted and measured crack speeds in a pressurized pipeline, results of Kanninen et al. [16].

The next step involved the development of an expression for the crack driving force G as a function of $\bar{w}$. Omitting the details, a relation $G = G(\dot{a})$ was obtained as a function of the pipe diameter and wall thickness; the elastic modulus, Poisson's ratio, and yield stress of the pipe material; the specific heat ratio, speed of sound, and pressure of the undisturbed gas; and the shear modulus of the soil surrounding the pipe. A typical result is shown in Fig. 5. Of particular importance is the existence of a maximum crack driving force for any given set of pipe geometry and operating conditions. This type of result was used to estimate the minimum crack arrest toughness value. Also of interest in this solution is that the limiting crack speed agrees with (9).

To assess the accuracy of the model, crack speeds were calculated for comparison with the experimental results. This was done using a value of the dynamic fracture energy requirement obtained from a drop-weight tear test (the best available alternative) in conjunction with curves typified by those of Fig. 5. In this way "the steady state", high speed limit was obtained. As shown in Fig. 6, the crack speeds so determined were found to compare well with observed speeds in full-scale pipeline tests. Note that, while other pipe fracture models have been developed, the model just described is so far the only one that offers a prediction of crack speeds that can be compared directly with experimental results.

Analysis of a thermal shock event

Jung and Kanninen [17] performed a series of run/arrest computations, both statically and dynamically, for the Oak Ridge National Laboratory (ORNL) thermal shock experiment TSE-5A conducted by Cheverton et al. [18]. Like the pipe problem, the immediate difficulty in any such calculation, then as now, is the lack of a proven $K_{ID} = K_{ID}(\dot{a}, T)$ relation for the A508 steel vessel material. This relation might be expected to have the general character given by (4). Hence, for heuristic purposes, Jung and Kanninen reasoned that the crack speeds in a thermal shock event might be such that $\dot{a} \ll V_l$ whereupon the speed-dependent term can be neglected. They further assumed that the existing K_{Ia} data were collected under conditions such that they provide a sufficiently close approximation to K_{IA}. On this basis, kinetic calculations of the four separate crack jump lengths observed in TSE-5A were performed to contrast the static calculations. The computational results and the experimental crack jump lengths are shown in Table 1.

The computational results of Jung and Kanninen provided in Table 1 were made using initial conditions that correspond to the experimentally-determined times of crack initiation events in ORNL experiment TSE-5A, but are otherwise independent of the results of that experiment. Thus, a general conclusion is illustrated by the comparisons shown in Table 1. Provided the calculations are made on a common basis, the kinetic approach will always predict a larger crack jump at arrest than will the static approch. Specifically, the kinetic results given in Table 1 obtained using the small specimen data account for roughly half of the discrepancy between the static prediction and the measured values. It might be noted that the more recent results obtained by Kanninen et al. [19] for a pressurized thermal shock event are in good accord with this conclusion.

The K_{Ia} and K_{ID} property values used in the computations shown in Table 1 were taken as the average of laboratory test specimen data. Subsequent static computations using K_{Ia} values derived from the vessel experiment, which constitute a lower bound to the laboratory specimen data, were found to agree well with the measured crack jump lengths. It can thus be concluded that, while dynamic phenomena are clearly present in a thermal shock event, their effect is within the scatter in the material fracture toughness values. Consequently, if a static prediction indicates that a reasonably large margin of safety exists (e.g., the remaining ligament after the final arrest is large compared to that

Table 1. Comparison of static and kinetic predictions of crack jump lengths with experimental measurements in ORNL thermal shock experiment TSE-5A [17,18]

Run/arrest event	Crack jump length (mm)		
	Static prediction	Kinetic prediction	Experimental measurement
First	2.1	6.6	5.8
Second	3.8	7.3	13.7
Third	6.1	11.0	10.7
Fourth	23.2	27.2	39.7
Total	35.2	52.1	69.9

which would produce a plastic instability), then no more rigorous computation is necessary. Of course, if the margin is narrow, the greater amount of penetration predicted by a kinetic approach could be significant so that the use of such computations is appropriate.

4. Nonlinear aspects of dynamic fracture mechanics

As should be apparent from the discussion to this point, applications of dynamic fracture mechanics for crack arrest are largely confined to elastodynamic considerations. Nonetheless, questions that are now outstanding appear to be resolvable only by resort to nonlinear dynamic treatments. One of these bears on the uniqueness of the K_{ID} property. Most investigators have found K_{ID} values that are reasonably geometry-independent, as they must be for K_{ID} to qualify as a material property. This is a key point because other investigations have revealed a small, but systematic, dependence on specimen geometry. A particularly revealing result is that of Kalthoff [20] shown in Fig. 7.

One explanation for a geometry dependence is that K_{ID} may not be a function only of the instantaneous crack speed but may also depend upon higher derivatives of crack length. An intuitive argument against this was given by Eshelby who stated that the "crack tip has no inertia", meaning that the acceleration of the crack tip does not affect the calculated value of the crack driving force. It is generally inferred from this that the K_{ID} property also cannot depend on the acceleration. It must therefore be concluded that local nonlinear effects must be responsible for whatever nonuniqueness in K_{ID} that exists.

The condition that the inelastic deformation surrounding the crack tip is "dominated" by the elastic K field sets a requirement for contained or small-scale yielding. But, because a moving crack inevitably leaves a wake of relaxed plasticity behind, except for very short crack jumps, LEFM clearly cannot in principle offer a valid characterization of a propagating crack. It might be noted that the same conceptual difficulty arises in fatigue, though at generally much lower stress levels, where it is dealt with by appealing to the idea of "similitude". For similitude the parameters governing the crack growth rate must be measured for the same type of load history as in the application of interest. When similitude is violated, the fatigue relations do not work, e.g., in crack growth retardation which follows an overload. One might therefore expect the same kind of effects in dynamic fracture. Hence, just as current advanced research efforts in fatigue are now being made via direct consideration of plastic yielding and relaxation, dynamic fracture mechanics researchers are also turning to elastic-plastic analyses.

A difficulty that immediately arises in elastic-plastic analyses is in identifying the proper crack growth criterion. A critical strain at a critical distance ahead of the crack tip has been proposed, for example. Freund and Douglas [21] have used this criterion effectively in deducing running fracture resistance values. But, such a parameter is

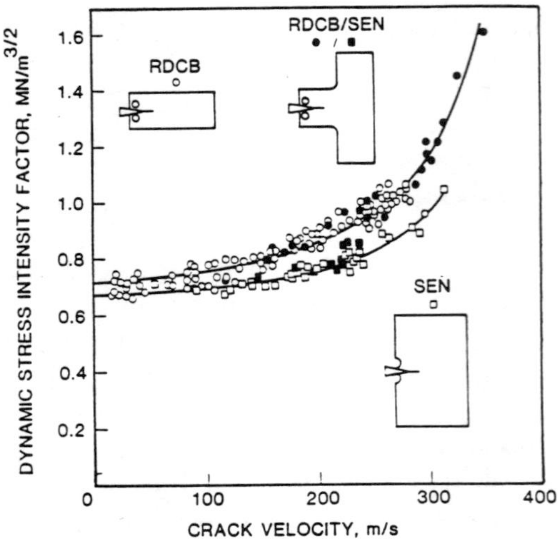

Figure 7. Geometry-dependence of the dynamic fracture toughness revealed by results in different specimens, results of Kalthoff [20].

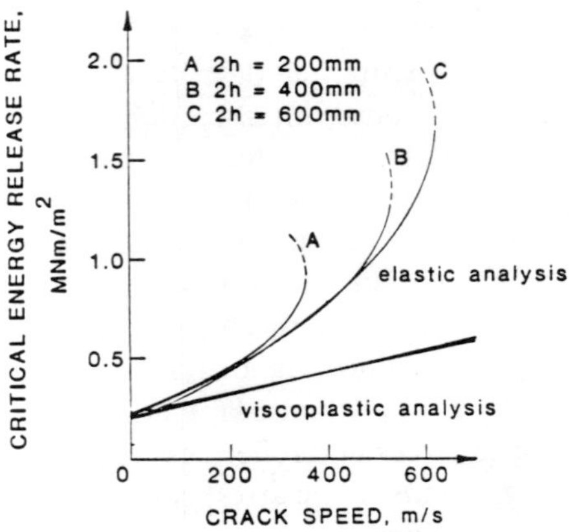

Figure 8.

somewhat unappealing – it cannot be measured and is not in any event palatable unless it can somehow be connected to a micromechanical picture. While results have been obtained in anti-plane strain that show a connection between ductile hole growth and a continuum strain measure ahead of a crack tip, a general result has yet to be obtained. An alternative to a critical strain criterion is the crack tip opening angle (CTOA). Use of this parameter to govern a fast propagating crack may have been first employed by Kanninen [22]. This parameter is attractive from both a computational point of view and from the extensive experience gained in elastic-plastic fracture mechanics that shows the constancy of the CTOA in stable growth. Nevertheless, the use of this parameter similarly requires a proper experimental and theoretical basis that does not now exist.

Except for the fact that a generally accepted crack advance criterion for such conditions does not exist, dynamic elastic-plastic and dynamic viscoplastic crack propagation computations pose no difficulty in principle. Important work on the characterization of the crack tip for dynamic elastic-plastic conditions has been given by Freund and Douglas [21], Achenbach et al. [23], Dantam and Hahn [24] and Ahmad et al. [25]. The essence of the problem is that a wake of relaxed plasticity is left behind the moving crack tip and this violates the "*K*-dominance" requirement of small-scale yielding that is a cornerstone of linear elastic fracture mechanics. Hence, an approach based upon incremental plasticity is required. But, owing to the high rates of deformation associated with a rapidly propagating crack, viscoplastic treatments that account for strain rate effects may well be required. A start on this type of problem has been made by Aboudi and Achenbach [26] and by Brickstad [27]. An illustrative example is given in Fig. 8 that reveals the practical effect of treating viscoplastic material behavior. These results were obtained by Brickstad [27] using the viscoplastic model of Perzyna in analyzing experimental test data on crack propagation in single edge notched (SEN) plates. The elastic interpretation of these results showed distinct geometry-dependence. But, by including viscoplastic effects, these have essentially been eliminated. Indeed, as Fig. 8 shows, much of the speed-dependence apparently is also removed in a viscoplastic treatment. Other treatments of this type are now in progress; for example, see Hudak et al. [28].

5. Discussion

The word "dynamic" has been used to connote crack growth processes accompanied by rapidly occurring changes in the crack/structure geometry and where these changes are not necessarily well described by a sequence of static equilibrium states. (The latter usage identifies those initial-boundary value problems where the inclusion of inertia forces is necessary.) This ambiguity has caused a great deal of unnecessary confusion. To avoid this, we have used the term "kinetic" to designate an analysis model that explicitly includes the propagating crack and "static" for those analyses that consider only the end points of such an event. Both the kinetic and the static models are of course part of dynamic fracture mechanics and, as the foregoing discussions indicated, both have their place in engineering analysis.

An elastodynamic kinetic analysis admits two complications beyond a static analysis: inertia forces and reflected stress waves. If K_{Ia} is the statically determined value of K that corresponds to the arrest condition, then, if there are no reflected stress waves, K_{Ia} will be exactly equal to K_{IA}. It follows that K_{Ia} is a perfectly legitimate fracture parameter provided that it is measured in a short jump crack arrest test; i.e., before reflected stress waves can impinge upon the crack tip. For larger crack jumps, K_{Ia} is still useful because it then provides a lower bound conservative estimate to the true crack arrest measure, K_{IA}.

It is commonly accepted that the use of the static crack arrest approach is conservative. Indeed, if one were to infer a value of K_{Ia} from a DCB test with a large crack jump, the true fracture toughness would clearly be underestimated. However, the use of such a value in an application to a structure via a static approch is not inherently conservative because the dynamically computed stress intensity factor can exceed its static counterpart. The results provided in the above for a thermal shock experiment indicate that dynamic effects enhance the crack driving force, although the magnitude of the effect is fairly modest for the conditions considered. In contrast, also as shown above, dynamic effects are very strong in a laboratory specimen such as a DCB specimen and cannot generally be ignored for such configurations.

6. Conclusions

While a significant dynamic increase in K does not always occur in an actual structure, it should always be recognized that such a possibility does exist. Unless it is certain that reflected stress waves cannot return to the crack tip, at least some initial kinetic calculations may be prudent in critical applications. This is particular true when crack arrest data are to be obtained via generation-phase analyses for cracks that do not arrest. Nevertheless, it is primarily in laboratory test specimens that dynamic effects are of most concern and, therefore, where kinetic analyses are of most value. It appears that many application areas exist where the greater simplicity of the static approach can be safely employed and, provided the crack arrest property is accurately obtained, where reliable margin of safety estimates can be expected from it.

Acknowledgement

This paper was prepared as part of a project in the dynamic fracture mechanics activity of the Office of Naval Research. The author would like to thank Dr. Yapa Rajapakse of ONR for his continued support in this area. Background material was also obtained through the author's participation in projects supported by the Heavy-Section Steel Technology Program of the Oak Ridge National Laboratory and by the Electric Power Research Institute. The author is indebted to Dr. Claud Pugh, ONRL, Mr. Milton Vagins, USNRC, and Dr. Tim Griesbach, EPRI, for these activities.

References

[1] M.F. Kanninen and C.H. Popelar, *Advanced Fracture Mechanics,* Oxford Press, New York (1985) to appear.
[2] G.R. Irwin and A.A. Wells, *Metallurgical Reviews* 10 (1965) 223–270.
[3] G.I. Barenblatt, *Advances in Applied Mechanics* 7 (1962) 55–129.
[4] P.B. Crosley and E.J. Ripling, *Journal of Basic Engineering* 91 (1969) 525–534.
[5] G.T. Hahn, R.G. Hoagland, M.F. Kanninen and A.R. Rosenfield, in *Dynamic Crack Propagation,* G.C. Sih, Noordhoff, Leiden, The Netherlands (1973) 649–662.
[6] J.F. Kalthoff, J. Beinert and S. Winkler, in *Fast Fracture and Crack Arrest,* G.T. Hahn and M.F. Kanninen, eds., ASTM STP 627, American Society Testing and Materials, Philadelphia (1977) 161–176.
[7] A.S. Kobayashi, K. Seo, J.Y. Jou and Y. Urabe, *Experimental Mechanics* 20 (1980) 73–79.
[8] M.F. Kanninen, in *Prospects of Fracture Mechanics,* G.C. Sih, et al., eds., Noordhoff International Publishing, Leyden, The Netherlands (1974) 251–266.
[9] P.B. Crosley, W.L. Fourney, G.T. Hahn, R.G. Hoagland, G.R. Irwin, and E.J. Ripling, *Cooperative Test Program on Crack Arrest Toughness Measurements.* NUREG/CR-3261, Nuclear Regulatory Commission, Washington, D.C. (April 1983).
[10] A.J. Rosakis, J. Duffy and L.B. Freund, in *Workshop on Dynamic Fracture,* California Institute of Technology, W.G. Knauss et al., eds., Pasadena, California (1983) 110–118.
[11] M.F. Kanninen, *International Journal of Fracture* 10 (1974) 415–430.
[12] J.E. Srawley and B. Gross, *Materials Research and Standards* 7 (1967) 155–162.
[13] W.B. Fichter, *International Journal of Fracture* 22 (1983) 133–143.
[14] W.A. Maxey, R.J. Eiber, R.J. Podlasek and A.R. Duffy, in *Crack Propagation in Pipelines,* Institute of Gas Engineers, London (1974).
[15] M.F. Kanninen, *International Journal of Fracture* 6 (1970) RCR 94–95.
[16] M.F. Kanninen, S.G. Sampath and C.H. Popelar, *Journal of Pressure Vessel Technology* 98 (1976) 56–65.
[17] J. Jung and M.F. Kanninen, *Journal of Pressure Vessel Technology* (1983) 111–116.
[18] R.D. Cheverton, O.A. Canonico, S.K. Iskander, S.E. Bolt, P.P. Holtz, R.K. Nanstad and W.J. Stelzman, *Journal of Pressure Vessel Technology* 105 (1983) 102–110.
[19] M.F. Kanninen, J. Ahmad, V. Papaspyropoulos, S.J. Hudak and J.W. FitzGerald, in *International Conference on Dynamic Fracture Mechanics*, San Antonio, Texas, 7–9 November 1984.
[20] J.F. Kalthoff, in *Workshop on Dynamic Fracture*, W.G. Knauss et al., eds. California Institute of Technology (1983) 11–35.
[21] L.B. Freund and A.S. Douglas, *Journal of the Mechanics and Physics of Solids* (1982) 59–74.
[22] M.F. Kanninen, *Journal of the Mechanics and Physics of Solids* 16 (1968) 215–228.
[23] J.D. Achenbach, M.F. Kanninen and C.H. Popelar, *Journal of the Mechanics and Physics of Solids* 29 (1981) 211–225.
[24] V. Dantam and G.T. Hahn, in *Fracture Tolerance Evaluation,* T. Kanazawa, A.S. Kobayashi, and K. Iida, eds., Toyoprint Co. Ltd., Japan (1982).
[25] J. Ahmad, J. Jung, C.R. Barnes and M.F. Kanninen, *Engineering Fracture Mechanics* 17 (1983) 235–246.
[26] J. Aboudi and J.D. Achenbach, *Engineering Fracture Mechanics* 18 (1983) 109–119.
[27] B. Brickstad, *Journal of the Mechanics and Physics of Solids* 31 (1983) 307–327.
[28] S.J. Hudak, K. Chan, K. Reed, J.H. FitzGerald, M.F. Kanninen, and J.D. Achenbach, in *International Conference on Dynamic Fracture Mechanics,* San Antonio, Texas, 7–9 November 1984.

Résumé

On étudie la compatibilité des approches statiques et dynamiques pour la propagation et l'arrêt d'une fissure. Pour des raisons de clarification, le terme "cinétique" est introduit dans une analyse qui couvre l'entièreté du processus d'amorçage, propagation et arrêt d'une fissure, par opposition à l'analyse statique qui ne considère que les points terminals et dans la mesure où les effets dynamiques ne sont pas toujours appréciables dans un processus de course et d'arrêt d'une fissure. On discute également l'importance d'intéger le travail expérimental et le travail par calculs dans ce domaine et l'on établit une différence entre les différentes voies qui peuvent être suivies. A titre d'exemple, on établit le calcul pour un problème de choc termique où les effets dynamiques sont minimes, c'est-à-dire où les approches statiques et cinétiques sont à peu près les mêmes et un calcul pour une éprouvette Cantilever où des différences significatives surviennent entre les deux approches. On conclut que les approches statiques et cinétiques sont entièrement compatibles pour autant que les ondes de contrainte réfléchies n'atteignent pas l'extrémité de la fissure avant que ne se produisent son arrêt. Cependant, comme lorsque cela n'est pas le cas, c'est l'approche cinétique qui doit être utilisée. De même, lorsque les effets inélastiques sont importants, seule l'approche cinétique peut les prendre en compte.